LABORATORY MANUAL FOR PULSE-WIDTH MODULATED DC–DC POWER CONVERTERS

LABORATORY MANUAL FOR PULSE-WIDTH MODULATED DC–DC POWER CONVERTERS

Marian K. Kazimierczuk and Agasthya Ayachit

Wright State University, Dayton, Ohio, USA

WILEY

Library of Congress Cataloging-in-Publication Data applied for.

A catalogue record for this book is available from the British Library.

ISBN: 9781119052760

Set in 10/12pt Times by Aptara Inc., New Delhi, India
Printed in Singapore by C.O.S. Printers Pte Ltd

1   2016

Contents

Part III   SEMICONDUCTOR MATERIALS AND POWER DEVICES

Preface

The *Laboratory Manual for Pulse-Width Modulated DC–DC Power Converters* is intended to aid undergraduate and graduate students of electrical engineering, practicing engineers, scientists, and circuit designers to have a grasp on designing and simulating a variety of fundamental and advanced power electronic circuits. The manual enables users to get accustomed to different simulation tools such as MATLAB®, Synopsys SABER®, LTSpice®, PLECS®, or any other Spice-based circuit simulation platforms. The approach presented in this manual will enhance a student's understanding of different power electronic converters and also gain knowledge in great depth on performing circuit simulations; a characteristic needed for a career in electrical engineering.

This manual is a supplementary material to the successful edition of the textbook *Pulse-width Modulated DC–DC Power Converters,* Second Edition, authored by Prof. Marian K. Kazimierczuk. The lab manual complements the content of the textbook and the combination of the two is a one-stop arrangement for students and instructors to gain the most about power electronic circuits and their simulation. This book features the following attributes:

  i. Unique in the market of textbooks for power electronics.
 ii. Can be adopted as a supplementary material for any commercially available textbooks on power electronics as well as classnotes.
iii. Can be used for distance-learning power electronic course or e-learning.
 iv. The software-oriented approach makes it convenient for students to have take-home assignments.
  v. Simple and easy-to-understand procedure set.
 vi. Provides a quick overview of various power converters and components.

The purpose of the *Laboratory Manual for Pulse-Width Modulated DC–DC Power Converters* is to provide a comprehensive instruction set for the following:

a. To design and simulate various topologies of power electronic dc–dc converters such as boost and boost-derived, buck and buck-derived, flyback, forward, half-bridge, and full-bridge converter topologies, operating in continuous-conduction mode or discontinuous-conduction mode.
b. To simulate the small-signal models of the power electronic circuits and to understand the different small-signal characteristics of boost and buck converters operating in continuous-conduction mode.

c. To understand the properties of silicon, silicon-carbide, and gallium-nitride power MOSFETs used in power electronic applications.

The topics presented in this lab manual have been thoughtfully considered, keeping in mind the benefits it offers to the students. The primary author of this lab manual has been teaching specialized graduate-level courses in power electronics for more than 25 years. Since then, consistent effort has been put into creating equally interesting and accurate lab curricula for the students. The outcome of that dedication has been this lab manual. The experiments in this manual have been tested and updated regularly for technical correctness and clarity in the presentation style. The authors of this book recommend the below instructions for instructors and students in making the best use of this manual.

For Instructors

Instructors involved in teaching power electronic courses can adopt this lab manual as a required course material. The lab manual consists of three parts:

> PART I—Open-Loop Pulse-Width Modulated DC–DC Power Converters
>
> PART II—Closed-Loop Pulse-Width Modulated DC–DC Power Converters
>
> PART III—Semiconductor Materials and Power Devices

Part I consists of 28 lab experiments. Part II has 16 lab experiments. Part III offers 6 lab experiments with several subsections. For an undergraduate power electronic course offered once an academic year, instructors can adopt selected lab topics from all of the three parts. For graduate programs offering specialized power electronic courses and taught for more than one semester, instructors may plan to dedicate a single part for every semester. The post-lab exercise will aid in summarizing the lab activity that was performed and provides a background for the consecutive labs. The students must be encouraged to follow the format of lab report at the end of the Appendix. Following such a format will aid in improving the students' technical communication and problem solving skills as well as their professional writing capabilities.

For Students

The lab manual assumes that the student is familiar with general circuit analysis techniques, electronic circuits, and the basic know-how of simulation tools. Every lab topic includes a pre-lab section. Students are encouraged to understand the circuit operation, calculate the component values, and also understand the design process of the converter under consideration. It is expected that the students follow every step in the procedure section for successful completion of the lab topic. The quick design section provides data about the component values and component selection that students can use for rapid verification. The Appendix has been made very resourceful and provides the following information:

a. Detailed summary of design equations for buck, boost, buck–boost, flyback, forward, half-bridge, and full-bridge converter in continuous-conduction mode and discontinuous-conduction mode.

b. Instructions on using MATLAB®, Synopsys SABER®, and other Spice-based simulation tools.
c. Spice models of several power diodes and power MOSFETs needed for component selection.
d. A summary of physical constants, values of different properties of silicon (Si), silicon-carbide (SiC), and gallium-nitride (GaN) semiconductor materials and power devices.
e. Format and guidelines to prepare a well-organized lab report.

Acknowledgments

Throughout the entire course of this project, the support provided by John Wiley and Sons, Ltd was exceptional. We wish to express our sincere thanks to Peter Mitchell, Publisher, Electrical Engineering; Ella Mitchell, Associate Commissioning Editor; and Liz Wingett, Project Editor for their cooperation. Our thanks are also to the team under Baljinder Kaur at Aptara Co. for their patience and tireless efforts in organizing the editing process. We would also like to extend our thanks and great appreciation to our families for their support.

Selected experiments in this lab manual were administered to power electronic graduate students at Wright State University. The results of these experiments performed by the students have been recorded continuously for better accuracy and improvement in the instruction set. We would like to thank the efforts of several students who were directly or indirectly involved in making this manuscript complete.

We have sincerely attempted at making this edition of the lab manual error-free so that students gain a better understanding of the course material. The authors would welcome and greatly appreciate readers' suggestions, corrections for improvements of the technical content as well as the presentation style, and ideas for newer topics, which can be implemented in possible future editions.

Marian K. Kazimierczuk
Agasthya Ayachit
Dayton, Ohio
USA

List of Symbols

A	Transfer function of forward path in negative feedback system
A_i	Inductor-to-load current transfer function
A_J	Cross-sectional area of junction
BW	Bandwidth
C	Filter capacitance
C_b	Blocking capacitance
C_c	Coupling capacitance
C_{ds}	Drain–source capacitance of MOSFET
C_{gd}	Gate–drain capacitance of MOSFET
C_{gs}	Gate–source capacitance of MOSFET
C_{iss}	MOSFET input capacitance at $V_{DS} = 0$, $C_{iss} = C_{gs} + C_{gd}$
C_{min}	Minimum value of filter capacitance C
C_o	Transistor output capacitance
C_{oss}	MOSFET output capacitance at $V_{GD} = 0$, $C_{oss} = C_{gs} + C_{ds}$
C_{ox}	Oxide capacitance per unit area
C_{rss}	MOSFET transfer capacitance, $C_{rss} = C_{gd}$
c	Speed of light
D	DC component of on-duty cycle of switch
d	AC component of on-duty cycle of switch
D_m	Amplitude of small-signal component of on-duty cycle of switch
d_T	Total on-duty cycle of switch
ESR	Equivalent series resistance of capacitors and inductors
f_c	Gain-crossover frequency
f_z	Frequency of zero of transfer function
f_0	Corner frequency
f_p	Frequency of pole of transfer function
f_s	Switching frequency
f_{-180}	Phase-crossover frequency
I_D	Average diode current
I_{DM}	Peak diode current
I_{Drms}	rms value of diode current
I_I	DC input current of converter
I_L	Average current through inductor L
I_{LB}	Average current through inductor L at CCM/DCM boundary

I_O DC output current of converter
I_{Omax} Maximum value of dc load current I_0
I_{Omin} Minimum value of dc load current I_0
I_{SM} Peak switch current
i_o AC component of load current
i_C Current through filter capacitor C
i_D Diode current
i_L Current through inductor L
i_O Total load current
i_S Switch current
k Boltzmann constant
L Inductance, Channel length
L_e Effective channel length
L_n Electron diffusion length
L_p Hole diffusion length
L_m Magnetizing inductance of transformer
L_{max} Maximum inductance L for DCM operation
L_{min} Minimum inductance L for CCM operation
M_{VDC} DC voltage transfer function of converter
M_v Open-loop input-to-output voltage function of converter
M_{vcl} Closed-loop input-to-output voltage function of converter
M_{vi} Open-loop input voltage-to-inductor current transfer function
M_{vo} Open-loop input-to-output voltage function of converter at $f = 0$
m_e Mass of free electron
m_e^* Effective mass of electron
m_h Mass of hole
m_e^* Effective mass of hole
N_A Concentration of acceptors
N_D Concentration of donors
N_p Number of turns of primary winding
N_s Number of turns of secondary winding
n Transformer turns ratio, electron concentration density
n_i Intrinsic carrier concentration
n_{pO} Thermal equilibrium minority electron concentration
p_{pO} Thermal equilibrium minority hole concentration
PM Phase margin
P_I DC input power of converter
P_{LS} Overall power dissipation of converter
PM Phase margin
P_O DC output power of converter
P_{RF} Conduction loss in diode forward resistance R_F
P_{rC} Conduction loss in filter capacitor ESR
P_{VF} Conduction loss in diode offset voltage V_F
p Hole concentration
Q Quality factor

Q_g	Gate charge
q	Magnitude of electron charge
r	Total parasitic resistance
R_{DR}	Resistance of drift region
R_F	Diode forward resistance
R_L	DC load resistance
R_{LB}	DC load resistance at CCM/DCM boundary
R_{Lmax}	Maximum value of load resistance R_L
R_{Lmin}	Minimum value of load resistance R_L
r_C	Equivalent series resistance (ESR) of filter capacitor
r_{DS}	On-resistance of MOSFET
q	Electron charge
S_{max}	Maximum percentage overshoot
T	Switching period, Loop gain
T_A	Ambient temperature
T_c	Voltage transfer function of controller
T_{cl}	Closed-loop control-to-output transfer function
T_J	Junction temperature
T_m	Transfer function of pulse-width modulator
T_p	Open-loop control-to-output transfer function
T_{po}	Open-loop control-to-output transfer function at $f = 0$
t_f	Fall time
t_r	Rise time
t_{rr}	Reverse recovery time
V_{bi}	Built-in potential
V_C	DC component of control voltage
V_{Cpp}	Peak-to-peak ripple voltage of the filter capacitance
V_E	DC component of error voltage
V_t	Gate-to-source threshold voltage
V_{BD}	Breakdown voltage
V_{BR}	Reverse blocking (breakdown) voltage
V_{DM}	Reverse peak voltage of diode
V_{DS}	Drain–source dc voltage of MOSFET
V_{DSS}	Drain–source breakdown voltage of MOSFETs
V_F	Diode offset voltage, dc component of feedback voltage
V_I	DC component of input voltage of converter
V_O	DC output voltage of converter
V_R	DC reference voltage
V_r	Peak-to-peak value of output ripple voltage
V_{rcpp}	Peak-to-peak ripple voltage across ESR
V_{SM}	Peak switch voltage
V_T	Thermal voltage
V_{Tm}	Peak ramp voltage of pulse-width modulator
v_C	Total control voltage
v_c	AC component of control voltage
v_{DS}	Drain–source voltage of MOSFET

v_E	Total error voltage
v_F	Total feedback voltage
v_e	AC component of error voltage
v_d	Average drift velocity
v_f	AC component of feedback voltage
v_L	Voltage across inductance L
v_i	AC component of converter input voltage
v_o	AC component of converter output voltage
v_{sat}	Saturation velocity of carriers
v_r	AC component of reference voltage
v_{rc}	Voltage across ESR of filter capacitor
v_{th}	Thermal velocity of electron
v_{sat}	Saturated average drift velocity
W	Channel width
W_C	Energy stored in capacitor
w_L	Energy stored in inductor
Z_i	Open-loop input impedance of converter
Z_o	Open-loop output impedance of converter
β	Transfer function of feedback network
Δi_L	Peak-to-peak of inductor ripple current
η	Efficiency of converter
θ	Thermal resistance, Mobility degradation coefficient
μ	Carrier mobility
μ_p	Mobility of holes
μ_n	Mobility of electrons
ξ	Damping ratio
ρ	Resistivity
σ	Conductivity, Damping factor
τ	Minority carrier lifetime, Time constant
τ_n	Electron lifetime
τ_p	Hole lifetime
ϕ	Phase of transfer function, Magnetic flux
ω	Angular frequency
ω_c	Unity-gain angular crossover frequency
ω_0	Corner angular frequency
ω_p	Angular frequency of simple pole
ω_z	Angular frequency of simple zero

Part I

Open-Loop Pulse-Width Modulated DC–DC Converters—Steady-State and Performance Analysis and Simulation of Converter Topologies

1

Boost DC–DC Converter in CCM—Steady-State Simulation

Objectives

The objectives of this lab are:

- To design a pulse-width modulated (PWM) boost dc–dc converter operating in continuous-conduction mode (CCM) for the design specifications provided.
- To simulate the boost converter on a circuit simulator and to analyze its characteristics in steady state.
- To determine the overall losses and the efficiency of the boost converter.

Specifications

The specifications of the boost converter are as provided in Table 1.1.

Pre-lab

For the specifications provided, find the values of all the components and parameters for the boost dc–dc converter operating in CCM using the relevant design equations provided in Table A.1 in Appendix A.

Quick Design

Choose:

$L = 20$ mH, ESR of the inductor $r_L = 2.1$ Ω, $R_{Lmin} = 1.778$ kΩ, $R_{Lmax} = 35.6$ kΩ, $D_{min} = 0.579$, $D_{nom} = 0.649$, $D_{max} = 0.714$, $C = 1$ μF, ESR of the capacitor $r_C = 1$ Ω.

Laboratory Manual for Pulse-Width Modulated DC–DC Power Converters, First Edition.
Marian K. Kazimierczuk and Agasthya Ayachit.
© 2016 John Wiley & Sons, Ltd. Published 2016 by John Wiley & Sons, Ltd.

Table 1.1 Parameters and their values

Parameter	Notation	Value
Minimum dc input voltage	V_{Imin}	127 V
Nominal dc input voltage	V_{Inom}	156 V
Maximum dc input voltage	V_{Imax}	187 V
DC output voltage	V_O	400 V
Switching frequency	f_s	100 kHz
Maximum output current	I_{Omax}	0.225 A
Minimum output current	I_{Omin}	5% of I_{Omax}
Output voltage ripple	V_r	$< 0.01 V_O$

MOSFET: International Rectifier IRF430 n-channel power MOSFET with $V_{DSS} = 500$ V, $I_{SM} = 4.5$ A, $r_{DS} = 1.8$ Ω at $T = 25°C$, $C_o = 135$ pF, and $V_t = 4$ V.

Diode: ON Semiconductor MUR1560 with $V_{RRM} = 600$ V, $I_F = 15$ A, $R_F = 17.1$ mΩ, and $V_F = 1.5$ V.

Procedure

A. Simulation of the Boost Converter and its Analysis in Steady State

1. Construct the circuit of the boost converter shown in Figure 1.1 on the circuit simulator. Name the nodes and components for convenience. Enter the values of all the components.
2. Initially, simulate the converter at $V_I = V_{Inom} = 156$ V, $R_L = R_{Lmin} = 1.778$ kΩ, and $D = D_{nom} = 0.649$. Connect a pulse voltage source in order to provide the gate-to-source voltage at the MOSFET gate and source terminals. Set `time period = 10 µs`, `duty cycle/width = 0.649`, and `amplitude = 12 V`. Let the `rise time` and `fall time` be equal to zero (optional).
3. Set simulation type to `transient analysis`. Set `end time = 10 ms` and `time step = 0.1 µs`. Run the simulation.
4. Plot the following waveforms after successful completion of the simulation. You may display the waveforms on different figure windows for better clarity.
   - Gate-to-source voltage v_{GS}, drain-to-source voltage v_{DS}, and diode voltage v_D.
   - Output voltage v_O, output current i_O, and output power p_O.
   - Inductor current i_L, diode current i_D, and MOSFET current i_S.
   Zoom in to display the steady-state region.

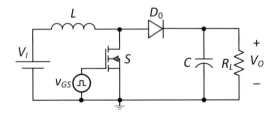

Figure 1.1 Circuit diagram of the PWM boost dc–dc converter.

5. Observe the inductor current waveform to ensure whether the current is in CCM. If the current is not in CCM, then increase the value of the inductor and repeat the simulation.
6. For the above-mentioned waveforms, measure:
   - The average and peak-to-peak values of the current through the inductor L.
   - The maximum and average values of the voltage across the MOSFET S.
   - The minimum and average values of the voltage across the diode D_0.
   - The maximum and average values of the currents through the MOSFET and the diode.
     Ensure that the values obtained above match the desired specifications. Should there be any deviation in the values, adjust the value of the duty cycle accordingly.
7. Repeat the steps above with duty cycle $D_{min} = 0.579$ and input voltage $V_{Imax} = 187$ V. Also, repeat for a duty cycle of $D_{max} = 0.714$ and an input voltage of $V_{Imin} = 127$ V.

B. Simulation of the Boost Converter to Determine the Power Losses and Overall Efficiency

1. Set up the converter to operate at the nominal operating condition, that is, at $D = D_{nom}$ and $V_I = V_{Inom}$. Let $R_L = R_{Lmin}$ such that the converter is delivering maximum output power.
2. Set the simulation type to transient analysis and perform the simulation.
3. Plot the waveforms of the input power p_I and the output power p_O. Zoom in to display the steady-state region.
4. Measure the average values of the input power and the output power. If the input power is negative, then consider only the magnitude of the average value of the input power.
5. Calculate the efficiency of the converter using $\eta = P_O/P_I$, where P_O is the average value of the output power and P_I is the average value of the input power, respectively.
6. This section may be repeated by plotting the power waveforms of all the components and then estimating their average values. All the power losses can be added to give the total power loss in the converter. Further, the efficiency can be estimated using $\eta = P_O/(P_{LS} + P_I)$, where P_O is the average value of the output power and P_{LS} is the sum of the average values of the power loss in individual components.
7. Repeat the above-mentioned activities in this section for $R_L = R_{Lmax}$ to determine the efficiency at minimum output power.

Post-lab Questions

1. Determine through simulations the maximum value of load resistance beyond which the converter operates in discontinuous-conduction mode?
2. Draw the waveforms of the inductor current for the input voltage at V_{Imin}, V_{Inom}, and V_{Imax}.
3. Draw the waveforms of the diode current, current through the filter capacitor, and current through the load resistor. Describe the nature of the three currents in terms of their shape and peak-to-peak values.
4. Draw the waveforms of the voltages across the capacitor and the equivalent series resistance of the capacitor.
5. Identify the component in the converter, which exhibits the highest power loss. Provide a proper reason for your answer.

2

Efficiency and DC Voltage Transfer Function of PWM Boost DC–DC Converter in CCM

Objectives

The objectives of this lab are:

- To design a PWM boost dc–dc converter in CCM for the given specifications.
- To analyze the variations in the efficiency of the lossy boost converter at different load resistances and input voltages.
- To determine the effect of duty cycle on the efficiency and steady-state dc voltage transfer function.

Theory

The steady-state dc voltage transfer function of a converter is

$$M_{VDC} = \frac{V_O}{V_I}. \tag{2.1}$$

The overall converter efficiency is

$$\eta = \frac{N_\eta}{D_\eta}, \tag{2.2}$$

Laboratory Manual for Pulse-Width Modulated DC–DC Power Converters, First Edition.
Marian K. Kazimierczuk and Agasthya Ayachit.
© 2016 John Wiley & Sons, Ltd. Published 2016 by John Wiley & Sons, Ltd.

where

$$N_\eta = R_L - M_{VDC}(R_F + r_C - r_{DS}) + \left\{ \left[M_{VDC}(R_F + r_C - r_{DS}) - R_L \right]^2 \right.$$

$$\left. - 4M_{VDC}^2 R_L(r_L + r_{DS}) \left(1 + \frac{V_F}{V_O} - \frac{r_C}{R_L} + f_s C_o R_L \right) \right\}^{\frac{1}{2}},$$

and

$$D_\eta = 2R_L \left(1 + \frac{V_F}{V_O} - \frac{r_C}{R_L} + f_s C_o R_L \right),$$

where r_L is the equivalent series resistance (ESR) of the boost inductor, r_{DS} is the on-state resistance of the MOSFET, D is the duty cycle, R_L is the load resistance, R_F is the forward resistance of the diode, r_C is the ESR of the capacitor, V_F is the forward voltage drop of the diode, V_O is the output voltage, f_s is the switching frequency, and C_o is the output capacitance of the MOSFET.

The steady-state dc voltage transfer function of the lossy boost converter is

$$M_{VDC} = \frac{\eta}{1 - D}. \tag{2.3}$$

Specifications

The specifications of the boost converter are as provided in Table 2.1.

Pre-lab

For the given specifications, find the values of all the components and parameters for the boost dc–dc converter operating in CCM using the design equations provided in Table A.1 in Appendix A.

Table 2.1 Parameters and their values

Parameter	Notation	Value
Minimum dc input voltage	V_{Imin}	127 V
Nominal dc input voltage	V_{Inom}	156 V
Maximum dc input voltage	V_{Imax}	187 V
DC output voltage	V_O	400 V
Switching frequency	f_s	100 kHz
Maximum output current	I_{Omax}	0.225 A
Minimum output current	I_{Omin}	5% of I_{Omax}
Output voltage ripple	V_r	$< 0.01 V_O$

Quick Design

Choose: $L = 20$ mH, ESR of the inductor $r_L = 2.1$ Ω, $R_{Lmin} = 1.778$ kΩ, $R_{Lmax} = 35.6$ kΩ, $D_{min} = 0.579$, $D_{nom} = 0.649$, $D_{max} = 0.714$, $C = 1$ μF, ESR of the capacitor $r_C = 1$ Ω.

MOSFET: International Rectifier IRF430 n-channel power MOSFET with $V_{DSS} = 500$ V, $I_{SM} = 4.5$ A, $r_{DS} = 1.8$ Ω at $T = 25\,°$C, $C_o = 135$ pF, and $V_t = 4$ V.

Diode: ON Semiconductor MUR1560 with $V_{RRM} = 600$ V, $I_F = 15$ A, $R_F = 17.1$ mΩ, and $V_F = 1.5$ V.

Procedure

A. Efficiency as a Function of the Input Voltage at Full and Light Load Conditions

1. The equation for the efficiency of the lossy boost converter is given in Equation (2.2). Define the equation for the efficiency on MATLAB® along with the given specifications and component values. Initially, let R_L be equal to $R_{Lmin} = V_O/I_{Omax}$.
2. Let the term M_{VDC} in Equation (2.2) be replaced with V_O/V_I as given in Equation (2.1).
3. Define the input voltage as a variable. Let the input voltage V_I vary from 120 to 190 V in steps of 1 V.
4. Using MATLAB's® built-in `plot` command, plot the efficiency as a function of the input voltage at $R_L = R_{Lmin}$.
5. Repeat the activity by replacing R_L with $R_{Lmax} = V_O/I_{Omin}$. Plot the two curves on the same figure window. Use `legend` or `text` commands to label the different curves clearly.

B. Efficiency as a Function of the Output Current at Minimum, Nominal, and Maximum Input Voltages

1. Use the code developed in Section A. In this section, fix the value of the input voltage V_I at V_{Imin}.
2. Define the load resistance as $R_L = V_O/I_O$. Let the output current I_O vary from 0 to 0.25 A in steps of 0.001 A.
3. Plot the efficiency as a function of the output current for $V_I = V_{Imin}$ using the `plot` command.
4. Repeat the above steps by changing the input voltage to V_{Inom} and then to V_{Imax}. Plot the efficiency curves for the three different input voltages on the same figure window. Use `legend` or `text` commands to label the different curves clearly.

C. DC Voltage Transfer Function as a Function of the Duty Cycle

1. Use the code for the efficiency as developed in Section A. Replace the term M_{VDC} in Equation (2.2) with $1/(1 - D)$.
2. Define the equation for the dc voltage transfer function of the lossy boost converter as given in Equation (2.3). Initially, let $R_L = R_{Lmin} = V_O/I_{Omax}$. Define a duty cycle range, that is, vary D from 0 to 1 in steps of 0.001.

3. Plot the lossy dc voltage transfer function M_{VDC} given in Equation (2.3) as a function of the duty cycle for $R_L = R_{Lmin}$ using the `plot` command.

4. Repeat the activity for $R_L = R_{Lmax} = V_O/I_{Omin}$. Plot the curves of the voltage transfer function for the two different load resistances on the same figure window. Use `legend` or `text` commands to label the different curves clearly.

Post-lab Questions

1. Draw the MATLAB® plots of efficiency versus duty cycle for the three different load resistances. Identify the values of the duty cycle on the three plots, where the efficiency begins to roll-off.

2. From Question 1, what is the useful range of duty cycle, where the efficiency and gain are maximum?

3. In Section C, determine the value of the duty cycle at which the dc voltage transfer function achieves a maximum value for $R_L = R_{Lnom}$.

4. Draw representative plots of the dc voltage transfer function of the ideal and lossy boost converters in CCM.

5. The plot of efficiency versus input voltage exhibits almost no change in efficiency for variation in the input voltage. Justify.

3

Boost DC–DC Converter in DCM—Steady-State Simulation

Objectives

The objectives of this lab are:

- To design a pulse-width modulated (PWM) boost dc–dc converter operating in discontinuous-conduction mode (DCM) for the design specifications provided.
- To simulate the converter and analyze its steady-state characteristics.
- To determine the overall losses and efficiency of the boost converter in DCM.

Specifications

The specifications of the boost converter are as provided in Table 3.1.

Pre-lab

For the given specifications, find the values of the components and the different parameters for the boost dc–dc converter operating in DCM using the design equations provided in Table B.1 in Appendix B.

Quick Design

Choose:

$L = 3.3$ µH, ESR of the inductor $r_L = 0.05$ Ω, $R_{Lmin} = 12$ Ω, $D_{min} = 0.165$, $D_{max} = 0.606$, $C = 150$ µF, ESR of the capacitor $r_C = 0.01$ Ω.

MOSFET: International Rectifier IRF142 n-channel power MOSFET with $V_{DSS} = 100$ V, $I_{SM} = 24$ A, $r_{DS} = 0.11$ Ω at $T = 25°C$, $C_o = 100$ pF, and $V_t = 4$ V.

Diode: Fairchild Semiconductor MBR1060 with $V_{RRM} = 600$ V, $I_F = 20$ A, $R_F = 25$ mΩ, and $V_F = 0.4$ V.

Laboratory Manual for Pulse-Width Modulated DC–DC Power Converters, First Edition.
Marian K. Kazimierczuk and Agasthya Ayachit.
© 2016 John Wiley & Sons, Ltd. Published 2016 by John Wiley & Sons, Ltd.

Table 3.1 Parameters and their values

Parameter	Notation	Value
Minimum dc input voltage	V_{Imin}	8 V
Maximum dc input voltage	V_{Imax}	18 V
DC output voltage	V_O	24 V
Switching frequency	f_s	100 kHz
Maximum output power	P_{Omax}	48 W
Minimum output power	P_{Omin}	0 W
Output voltage ripple	V_r	$< 0.01 V_O$

Procedure

A. Simulation of the Boost Converter and its Analysis in Steady State

1. Construct the circuit of the boost converter shown in Figure 3.1 on the circuit simulator. Name all the nodes and components for convenience.
2. Initially, let the input voltage V_I be set to $V_{Imax} = 18$ V. The corresponding duty cycle is $D_{min} = 0.165$.
3. Enter the values of all the components. Connect a pulse voltage source in order to provide the gate-to-source voltage at the MOSFET gate and source terminals. Set `time period = 10 µs`, `duty cycle/width = 0.165`, and `amplitude = 12 V`. Let the `rise time` and `fall time` be equal to zero (optional).
4. Choose simulation type as `transient analysis`. Set `end time = 10 ms` and `time step = 0.1 µs`. Run the simulation.
5. Plot the following parameters after successful completion of the simulation. You may display the waveforms on different figure windows for better clarity.
   - Gate-to-source voltage v_{GS}, drain-to-source voltage v_{DS}, and diode voltage v_D.
   - Output voltage v_O, output current i_O, and output power p_O.
   - Inductor current i_L, diode current i_D, and MOSFET current i_S.
   Use the `zoom` option to display only the steady-state region.
6. Observe the inductor current waveform to ensure whether the current is in DCM. If the inductor current is not in DCM, then decrease the value of the inductor and repeat the simulation.
7. For the above-mentioned waveforms, measure:
   - The average and peak-to-peak values of the current through the inductor L.
   - The maximum, intermediate, and average values of the voltage across the MOSFET S.
   - The minimum, intermediate, and average values of the voltage across the diode D_0.

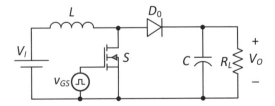

Figure 3.1 Circuit diagram of the PWM boost dc–dc converter.

- The maximum and average values of the currents through the MOSFET and the diode. Ensure that the values obtained above match the desired specifications.

8. Repeat all of the above steps by changing the values of the duty cycle D of the gate-to-source voltage source to $D_{max} = 0.606$ and the input voltage source to $V_{Imin} = 8$ V.

B. Simulation of the Boost Converter to Determine the Power Losses and Overall Efficiency

1. Set up the converter to operate at maximum input voltage V_{Imax} and at minimum duty cycle D_{min}. Let $R_L = R_{Lmin}$ such that the converter is operating at maximum output power.
2. Set the simulation type to transient analysis and perform the simulation.
3. Plot the waveforms of the input power p_I and the output power p_O. Zoom in to the steady-state region.
4. Measure the average values of the input power and the output power. If the input power is negative, then consider only the magnitude of the average value of the input power.
5. Calculate the efficiency of the converter using $\eta = P_O/P_I$, where P_O is the average value of the output power and P_I is the average value of the input power, respectively.
6. This section may be repeated by plotting the power waveforms of all the components and then estimating their average values. All the power losses can be added to give the total power loss in the converter. Further, the efficiency can be estimated using $\eta = P_O/(P_{LS} + P_I)$, where P_O is the average value of the output power and P_{LS} is the sum of the average values of the power loss in individual components.
7. Repeat the above-mentioned activities in this section for $V_I = V_{Imin}$. Adjust the duty cycle accordingly.

Post-lab Questions

1. Prepare a well-organized table showing the values of voltages and currents obtained in Section A.
2. In Section B, what is the difference in the efficiencies of the boost converter between its operation at maximum input voltage and minimum input voltage. What is the reason for such large variation in the efficiency?
3. Draw the waveform of the inductor current in DCM and label the slopes and axes details clearly.
4. How do you expect the efficiency to vary, when the load resistance is changed from no load to full load?

4

Efficiency and DC Voltage Transfer Function of PWM Boost DC–DC Converter in DCM

Objectives

The objectives of this lab are:

- To design a PWM boost dc–dc converter in DCM for the given specifications.
- To analyze the variations in the efficiency of the boost converter in DCM at different load resistances and different input voltages.
- To analyze the variation in the dc voltage transfer function of the lossy boost converter as a function of the duty cycle.

Theory

The steady-state dc voltage transfer function of a converter is

$$M_{VDC} = \frac{V_O}{V_I}. \tag{4.1}$$

In terms of the circuit parameters, the transfer function under ideal conditions can be expressed as

$$M_{VDC} = \frac{1 + \sqrt{1 + \frac{2D^2 V_O}{f_s L I_O}}}{2}, \tag{4.2}$$

Laboratory Manual for Pulse-Width Modulated DC–DC Power Converters, First Edition.
Marian K. Kazimierczuk and Agasthya Ayachit.
© 2016 John Wiley & Sons, Ltd. Published 2016 by John Wiley & Sons, Ltd.

resulting in

$$D = \sqrt{\frac{2f_s L I_O M_{VDC}(M_{VDC} - 1)}{V_O}}. \tag{4.3}$$

The overall converter efficiency is

$$\eta = \frac{1}{1 + \frac{P_{LS}}{P_O}}, \tag{4.4}$$

where P_{LS} is the total power loss in the converter

$$P_{LS} = \left[\frac{2r_{DS}}{3} \sqrt{\frac{2(M_{VDC} - 1)^3}{f_s L R_L M_{VDC}}} + \frac{2R_F}{3} \sqrt{\frac{2(M_{VDC} - 1)}{f_s L M_{VDC} R_L}} \right.$$
$$\left. + \frac{2r_L}{3} \sqrt{\frac{2M_{VDC}(M_{VDC} - 1)}{f_s L R_L}} + \frac{V_F}{V_O} + f_s C_o R_L \right] P_O, \tag{4.5}$$

where r_L is the equivalent series resistance (ESR) of the boost inductor, r_{DS} is the on-state resistance of the MOSFET, D is the duty cycle, R_L is the load resistance, R_F is the forward resistance of the diode, r_C is the ESR of the capacitor, V_F is the forward voltage drop of the diode, V_O is the output voltage, f_s is the switching frequency, and C_o is the output capacitance of the MOSFET. In terms of the duty cycle, the total power loss P_{LS} can be expressed as

$$P_{LS} = \left[\frac{4r_{DS} R_L D^3}{3f_s^2 L^2} \frac{1}{\left(1 + \sqrt{1 + \frac{2D^2 R_L}{f_s L}}\right)^2} + \frac{4DR_F}{3f_s L} \frac{1}{\left(1 + \sqrt{1 + \frac{2D^2 R_L}{f_s L}}\right)^2} \right.$$
$$\left. + \frac{2Dr_L}{3f_s L} + \frac{V_F}{V_O} + f_s C_o R_L \right] P_O. \tag{4.6}$$

The steady-state dc voltage transfer function of the lossy boost converter in DCM is

$$M_{VDC} = \frac{1 + \sqrt{1 + \frac{2\eta D^2 V_O}{f_s L I_O}}}{2} = \frac{1 + \sqrt{1 + \frac{2\eta D^2 R_L}{f_s L}}}{2}. \tag{4.7}$$

Specifications

The specifications of the boost converter are as provided in Table 4.1.

Pre-lab

For the given specifications, find the values of the components and the different parameters for the boost dc–dc converter operating in DCM using the design equations provided in Table B.1 in Appendix B.

Table 4.1   Parameters and their values

Parameter	Notation	Value
Minimum dc input voltage	V_{Imin}	8 V
Maximum dc input voltage	V_{Imax}	18 V
DC output voltage	V_O	24 V
Switching frequency	f_s	100 kHz
Maximum output power	P_{Omax}	48 W
Minimum output power	P_{Omin}	0 W
Output voltage ripple	V_r	$< 0.01 V_O$

Quick Design

Choose:

$L = 3.3$ µH, ESR of the inductor $r_L = 0.05$ Ω, $R_{Lmin} = 12$ Ω, $D_{min} = 0.165$, $D_{max} = 0.606$, $C = 150$ µF, ESR of the capacitor $r_C = 0.01$ Ω.

MOSFET: International Rectifier IRF142 n-channel power MOSFET with $V_{DSS} = 100$ V, $I_{SM} = 24$ A, $r_{DS} = 0.11$ Ω at $T = 25°$C, $C_o = 100$ pF, and $V_t = 4$ V.

Diode: Fairchild Semiconductor MBR1060 with $V_{RRM} = 600$ V, $I_F = 20$ A, $R_F = 25$ mΩ, and $V_F = 0.4$ V.

Procedure

A. Efficiency as a Function of the Input Voltage at Various Load Conditions

1. The equation for the efficiency of the lossy boost converter is given in Equation (4.4) and the expression for the total power loss is provided in Equation (4.5). Define the equation for the efficiency on MATLAB® along with given specifications and component values. Initially, let R_L be equal to $R_{Lmin} = 12$ Ω.
2. Let the term M_{VDC} in Equation (4.5) be replaced with V_O/V_I as given in Equation (4.1).
3. Define the input voltage as a variable. Let the input voltage V_I vary from 8 to 18 V in steps of 0.01 V.
4. Using MATLAB's® built-in `plot` command, plot the efficiency as a function of the input voltage at $R_L = R_{Lmin}$.
5. Repeat the activity for $R_L = 24$ Ω and $R_L = 48$ Ω. Plot the three curves on the same figure window. Use `legend` or `text` commands to label the different curves clearly.

B. Efficiency as a Function of the Output Current at Minimum, Nominal, and Maximum Input Voltages

1. Use the code developed in Section A. In this Section, fix the value of the input voltage V_I at V_{Imin}.
2. Define the load resistance as $R_L = V_O/I_O$. Let the output current I_O vary from 0 to 0.25 A in steps of 0.001 A.

3. Plot the efficiency as a function of the output current for $V_I = V_{Imin}$ using the `plot` command.

4. Repeat the above steps by changing the input voltage to V_{Inom} and then to V_{Imax}. Plot the efficiency curves for the three different input voltages on the same figure window. Use `legend` or `text` commands to label the different curves clearly.

C. DC Voltage Transfer Function as a Function of the Duty Cycle

1. The equation for the total power loss P_{LS} in the boost converter in terms of duty cycle D is given in Equation (4.6). Redefine the equation for efficiency as given in Equation (4.4) using Equation (4.6). Initially, let $R_L = R_{Lmin} = 12\ \Omega$.

2. Next, define the equation for the dc voltage transfer function of the lossy boost converter as given in Equation (4.7). Let the range of the duty cycle D be from 0 to 1 and provide a step size of 0.001.

3. Plot the lossy dc voltage transfer function M_{VDC} as a function of the duty cycle D for $R_L = R_{Lmin}$ using the `plot` command.

4. Repeat the activity for $R_L = 40\ \Omega$ and $R_L = 75\ \Omega$. Plot the curves of the voltage transfer function for the three different load resistances on the same figure window. Use `legend` or `text` commands to label the different curves clearly.

Post-lab Questions

1. Using Equation (4.6), identify the component with the highest value of power loss at D_{min} and R_{Lmin}.

2. Determine the conditions for the boost converter to operate at the boundary between continuous- and discontinuous-conduction modes.

5

Open-Loop Boost AC–DC Power Factor Corrector—Steady-State Simulation

Objectives

The objectives of this lab are:

- To simulate the boost ac–dc power factor correction circuit.
- To simulate the circuit of the boost converter operating as a peak rectifier.

Specifications

The specifications of the boost converter are as provided in Table 5.1. The input voltage to the boost converter is supplied by an ac voltage source having a nominal amplitude of $V_m = 110$ V and at a frequency of $f = 60$ Hz. In a practical set-up, the rms voltage of the US utility line changes from 90 (low line) to 132 V (high line) for normal operation. Hence, the minimum, nominal, and maximum values of the dc voltages at the output of a full-bridge front-end rectifier are V_{Imin}, V_{Inom}, and V_{Imax}.

Pre-lab

For the specifications provided, find the values of all the components and parameters for the boost dc–dc converter operating in CCM using the relevant design equations provided in Table A.1 in Appendix A.

Laboratory Manual for Pulse-Width Modulated DC–DC Power Converters, First Edition.
Marian K. Kazimierczuk and Agasthya Ayachit.
© 2016 John Wiley & Sons, Ltd. Published 2016 by John Wiley & Sons, Ltd.

Table 5.1   Parameters and their values

Parameter	Notation	Value
Minimum dc input voltage	V_{Imin}	127 V
Nominal dc input voltage	V_{Inom}	156 V
Maximum dc input voltage	V_{Imax}	187 V
DC output voltage	V_O	400 V
Switching frequency	f_s	100 kHz
Maximum output current	I_{Omax}	0.225 A
Minimum output current	I_{Omin}	5% of I_{Omax}
Output voltage ripple	V_r	$< 0.01 V_O$

Quick Design

Choose: $L = 20$ mH, ESR of the inductor $r_L = 2.1$ Ω, $R_{Lmin} = 1.778$ kΩ, $R_{Lmax} = 35.6$ kΩ, $D_{min} = 0.579, D_{nom} = 0.649, D_{max} = 0.714, C = 1$ μF, ESR of the capacitor $r_C = 1$ Ω. Choose the value of the bulk capacitor as $C_B = 1$ mF.

MOSFET: International Rectifier IRF430 n-channel power MOSFET with $V_{DSS} = 500$ V, $I_{SM} = 4.5$ A, $r_{DS} = 1.8$ Ω at $T = 25°C$, $C_o = 135$ pF, and $V_t = 4$ V.

Diode: ON Semiconductor MUR1560 with $V_{RRM} = 600$ V, $I_F = 15$ A, $R_F = 17.1$ mΩ, and $V_F = 1.5$ V.

Full-bridge rectifier: Use MUR1560 or ideal diodes in this stage.

Procedure

A. Simulation of the Boost Converter as a Power Factor Corrector

1. Construct the circuit of the boost power factor corrector as shown in Figure 5.1. Ensure that the bulk capacitor C_B is removed from the circuit.
2. Consider the nominal input voltage for the simulation, that is, let the amplitude of the ac sinusoidal voltage be $V_m = 110$ V at a frequency of $f = 60$ Hz. Let the value of the load resistance R_L be equal to $R_{Lmin} = 1.778$ kΩ.
3. Provide a pulse voltage source at the MOSFET's gate and source terminals. Set `time period` = 10 μs, `duty cycle/width` = 0.649, and `amplitude` = 12 v. Let the `rise time` and `fall time` be equal to zero (optional).

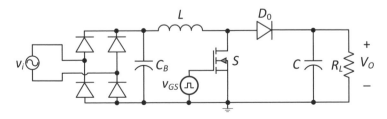

Figure 5.1   Circuit diagram of the PWM boost ac–dc converter.

4. Set simulation type to `transient analysis`. Set `end time = 40 ms` and `time step = 1 μs`. Run the simulation. An end time of 40 ms is essential in order to obtain at least two complete cycles of the input ac voltage.
5. Plot the following waveforms after successful completion of the simulation. You may display the waveforms on different figure windows for better clarity.
   - Input ac voltage, rectified ac voltage (or output of the rectifier).
   - Input ac current (before rectifier), inductor current i_L of the boost converter.
   - Gate-to-source voltage v_{GS}, drain-to-source voltage v_{DS}, and diode voltage v_D.
   - Output voltage v_O, output current i_O, and output power p_O.

   Superimpose the waveforms of the inductor current and the rectifier output voltage. Observe whether the two quantities follow each other in phase and magnitude. Ensure that the maximum value of the output voltage of the boost converter must be equal to V_O.

B. Simulation of the Boost Converter as a Peak Rectifier Circuit

1. Construct the circuit of the boost converter as peak rectifier as shown in Figure 5.1. Ensure that the bulk capacitor C_B is included in the circuit.
2. Repeat all the activities mentioned in Section A.
3. Ensure that the output voltage of the boost converter has a constant dc value of $V_O = 400$ V.

Post-lab Questions

1. Explain the term *power factor correction*.
2. Draw the waveforms of currents and voltages for a rectifier with (a) resistive load, (b) inductive load, and (c) capacitive load. Indicate what type of load presents the highest power factor.
3. From waveforms in Section A, determine the nature of the load presented to the rectifier circuit by the boost converter.
4. What is the difference in terms of applications, between the boost converter as power factor corrector and as a peak rectifier?

6

Buck DC–DC Converter in CCM—Steady-State Simulation

Objectives

The objectives of this lab are:

- To design a pulse-width modulated buck dc–dc converter operating in continuous-conduction mode (CCM) for the design specifications provided.
- To simulate and analyze the characteristics of the converter operating in steady state.
- To determine the overall losses and efficiency of the buck converter in CCM.

Specifications

The specifications of the buck converter are as provided in Table 6.1.

Pre-lab

Design the buck converter for the specifications provided above. The design equations needed to obtain the values of the components are provided in Table A.1 in Appendix A.

Quick Design

Assume an overall efficiency of $\eta = 85\%$. Choose:

$L = 40$ μH, ESR of the inductor $r_L = 0.05$ Ω, $R_{Lmin} = 1.2$ Ω, $R_{Lmax} = 12$ Ω, $D_{min} = 0.441$, $D_{nom} = 0.506$, $D_{max} = 0.588$, $C = 100$ μF, ESR of the capacitor $r_C = 50$ mΩ.

MOSFET: International Rectifier IRF150 n-channel power MOSFET with $V_{DSS} = 100$ V, $I_{SM} = 40$ A, $r_{DS} = 55$ mΩ at $T = 25°$C, $C_o = 100$ pF, and $V_t = 4$ V.

Diode: ON Semiconductor MUR1060 with $V_{RRM} = 60$ V, $I_F = 20$ A, $R_F = 25$ mΩ, and $V_F = 0.4$ V.

Laboratory Manual for Pulse-Width Modulated DC–DC Power Converters, First Edition.
Marian K. Kazimierczuk and Agasthya Ayachit.
© 2016 John Wiley & Sons, Ltd. Published 2016 by John Wiley & Sons, Ltd.

Table 6.1 Parameters and their values

Parameter	Notation	Value
Minimum dc input voltage	V_{Imin}	24 V
Nominal dc input voltage	V_{Inom}	28 V
Maximum dc input voltage	V_{Imax}	32 V
DC output voltage	V_O	12 V
Switching frequency	f_s	100 kHz
Maximum output current	I_{Omax}	10 A
Minimum output current	I_{Omin}	1 A
Output voltage ripple	V_r	$< 0.01 V_O$

Procedure

A. Simulation of the Buck Converter and its Analysis in Steady State

1. Construct the circuit of the buck converter shown in Figure 6.1 on the circuit simulator. Name all the nodes and components for convenience.
2. Initially, simulate the converter at $V_I = V_{Inom} = 28$ V, $R_L = R_{Lmin} = 1.2$ Ω, and $D = D_{nom} = 0.506$. Connect a pulse voltage source in order to provide the gate-to-source voltage at the MOSFET gate and source terminals. Set `time period = 10` μs, `duty cycle/width = 0.506`, and `amplitude = 12` V. Let the `rise time` and `fall time` be equal to zero (optional).
3. Set simulation type to `transient analysis`. Set `end time = 10` ms and a `time step = 0.1` μs. Run the simulation.
4. Plot the following parameters after successful completion of the simulation. You may display the waveforms on different figure windows for better clarity.
   - Gate-to-source voltage v_{GS}, drain-to-source voltage v_{DS}, and diode voltage v_D.
   - Output voltage v_O, output current i_O, and output power p_O.
   - Inductor current i_L, diode current i_D, and MOSFET current i_S.
   Zoom in to display only the steady-state region.
5. Observe the inductor current waveform to ensure whether the current is in CCM. If the current is not in CCM, then increase the value of the inductor and repeat the simulation.
6. For the above-mentioned waveforms, measure:
   - The average and peak-to-peak values of the current through the inductor L.
   - The maximum and average values of the voltage across the MOSFET S.
   - The minimum and average values of the voltage across the diode D_0.
   - The maximum and average values of the currents through the MOSFET and the diode.
   Ensure that the values obtained above match the desired specifications.

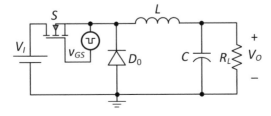

Figure 6.1 Circuit diagram of the PWM buck dc–dc converter.

7. Repeat the steps above with an input voltage of $V_{Imax} = 32$ V and a corresponding duty cycle of $D_{min} = 0.441$.
8. Next, repeat the activities for an input voltage of $V_{Imin} = 24$ V and a duty cycle of $D = 0.588$.

B. Simulation of the Buck Converter to Determine the Power Losses and Overall Efficiency

1. Set up the converter to operate at nominal operating condition, that is, at $D = D_{nom}$ and $V_I = V_{Inom}$. Let $R_L = R_{Lmin}$ such that the converter delivers a maximum output power.
2. Set the simulation type to transient analysis and perform the simulation.
3. Plot the waveforms of the input power p_I and the output power p_O. Zoom in to the steady-state region.
4. Measure the average values of the input power and the output power. If the input power is negative, then consider only the magnitude of the average value of the input power.
5. Calculate the efficiency of the converter using $\eta = P_O/P_I$, where P_O is the average value of the output power and P_I is the average value of the input power, respectively.
6. This section may be repeated by plotting the power waveforms of all the components and then estimating their average values. All the power losses can be added to give the total power loss in the converter. Further, the efficiency can be estimated using $\eta = P_O/(P_{LS} + P_I)$, where P_O is the average value of the output power and P_{LS} is the sum of the average values of the power loss in individual components.
7. Repeat the above-mentioned activities in this section for $R_L = R_{Lmax}$ to determine the efficiency at minimum output power.

Post-lab Questions

1. Represent the values obtained above in the form of a well-organized table.
2. Sketch the waveforms of the inductor current, current through the filter capacitor branch, and current through the load resistance. Describe the nature of the current through the equivalent series resistance (ESR) of the capacitor. What is the effect on the shape of the current when the value of the ESR is increased?
3. Explain how the switching frequency component is eliminated by the output filter network.
4. In Section B, identify the component that exhibits the highest power loss.

   Additional Activity: Add an inductor $L_c = 100$ nH in series with the filter capacitor. Repeat the activities in Section A. Observe the shape of the output voltage waveform, current through the filter branch, and voltages across all the components in the filter branch. The inductor L_c represents the equivalent series inductance or the parasitic inductance due to the metal leads of the filter capacitor.

7

Efficiency and DC Voltage Transfer Function of PWM Buck DC–DC Converter in CCM

Objectives

The objectives of this lab are:

- To design a PWM buck dc–dc converter in CCM for the design specifications provided.
- To analyze the variation in efficiency of the lossy buck converter at different load resistances and different input voltages.
- To analyze the variation in the lossy steady-state, dc voltage transfer function as a function of the duty cycle.

Theory

The steady-state dc voltage transfer function of a converter is

$$M_{VDC} = \frac{V_O}{V_I}. \tag{7.1}$$

The overall converter efficiency is

$$\eta = \frac{N_\eta}{D_\eta}, \tag{7.2}$$

Laboratory Manual for Pulse-Width Modulated DC–DC Power Converters, First Edition.
Marian K. Kazimierczuk and Agasthya Ayachit.
© 2016 John Wiley & Sons, Ltd. Published 2016 by John Wiley & Sons, Ltd.

where

$$N_\eta = 1 + M_{VDC}\left(\frac{V_F}{V_O} + \frac{r_C R_L}{6f_s^2 L^2} - \frac{r_{DS} - R_F}{R_L}\right) + \left\{\left[1 + M_{VDC}\left(\frac{V_F}{V_O} + \frac{r_C R_L}{6f_s^2 L^2} - \frac{r_{DS} - R_F}{R_L}\right)\right]^2\right.$$
$$\left. - \frac{M_{VDC}^2 r_C R_L}{3f_s^2 L^2}\left(1 + \frac{R_F + r_L}{R_L} + \frac{V_F}{V_O} + \frac{f_s C_o R_L}{M_{VDC}^2} + \frac{r_C R_L}{12f_s^2 L^2}\right)\right\}^{\frac{1}{2}},$$

and

$$D_\eta = 2\left(1 + \frac{R_F + r_L}{R_L} + \frac{V_F}{V_O} + \frac{f_s C_o R_L}{M_{VDC}^2} + \frac{r_C R_L}{12f_s^2 L^2}\right),$$

where r_L is the equivalent series resistance (ESR) of the buck inductor, r_{DS} is the on-state resistance of the MOSFET, D is the duty cycle, R_L is the load resistance, R_F is the forward resistance of the diode, r_C is the ESR of the capacitor, V_F is the forward voltage drop of the diode, V_O is the output voltage, f_s is the switching frequency, and C_o is the output capacitance of the MOSFET.

The steady-state dc voltage transfer function of the lossy buck converter is

$$M_{VDC} = \eta D. \tag{7.5}$$

Specifications

The specifications of the buck converter in CCM are as provided in Table 7.1.

Pre-lab

For the above specifications, determine the values of all the components and parameters for the buck dc–dc converter operating in CCM using the relevant design equations provided in Table A.1 in Appendix A.

Table 7.1 Parameters and their values

Parameter	Notation	Value
Minimum dc input voltage	V_{Imin}	24 V
Nominal dc input voltage	V_{Inom}	28 V
Maximum dc input voltage	V_{Imax}	32 V
DC output voltage	V_O	12 V
Switching frequency	f_s	100 kHz
Maximum output current	I_{Omax}	10 A
Minimum output current	I_{Omin}	1 A
Output voltage ripple	V_r	$< 0.01 V_O$

Quick Design

Assume an overall efficiency of $\eta = 85\%$. Choose:

$L = 40$ μH, ESR of the inductor $r_L = 0.05$ Ω, $R_{Lmin} = 1.2$ Ω, $R_{Lmax} = 12$ Ω, $D_{min} = 0.441$, $D_{nom} = 0.506$, $D_{max} = 0.588$, $C = 100$ μF, ESR of the capacitor $r_C = 50$ mΩ.

MOSFET: International Rectifier IRF150 n-channel power MOSFET with $V_{DSS} = 100$ V, $I_{SM} = 40$ A, $r_{DS} = 55$ mΩ at $T = 25°C$, $C_o = 100$ pF, and $V_t = 4$ V.

Diode: ON Semiconductor MUR1060 with $V_{RRM} = 60$ V, $I_F = 20$ A, $R_F = 25$ mΩ, and $V_F = 0.4$ V.

Procedure

A. Efficiency of the Buck Converter as a Function of the Input Voltage at Full and Light Load Conditions

1. Define the equation for the efficiency on MATLAB® as given in Equation (7.2) on MATLAB®. Also, define all the specifications and component values.
2. Replace the term M_{VDC} in the equation of efficiency with V_O/V_I. Define a range of input voltage, that is, vary V_I from 20 to 40 V in steps of 0.01 V. Initially, let $R_L = R_{Lmin} = V_O/I_{Omax}$.
3. Using MATLAB's® built-in `plot` command, plot the efficiency as a function of the input voltage at $R_L = R_{Lmin}$.
4. Next, repeat the activity by replacing R_L with $R_{Lmax} = V_O/I_{Omin}$. Plot the two curves on the same figure window. Use `legend` or `text` commands to label the different curves clearly.

B. Efficiency of the Buck Converter as a Function of the Output Current at Minimum, Nominal, and Maximum Input Voltages

1. Use the code developed in the previous section. Assign a fixed value for the input voltage. Let $V_I = V_{Imin}$.
2. Replace the term R_L with V_O/I_O. Define a range for the output current, that is, vary I_O from 0 to 10 A in steps of 0.001 A.
3. Plot the curve of efficiency as a function of the output current at $V_I = V_{Imin}$.
4. Next, repeat the result by replacing the input voltage with V_{Inom} and then with V_{Imax}. The three curves must be plotted on the same figure window. Use `legend` or `text` commands to label the different curves clearly.

C. DC Voltage Transfer Function of the Buck Converter as a Function of the Duty Cycle

1. Use the code for the efficiency as developed in Section A. Replace the term M_{VDC} in Equation (7.2) with D.
2. Using the equation for efficiency as given in Equation (7.2), define the expression for the dc voltage transfer function of the lossy buck converter as given in Equation (7.5). Initially,

let $R_L = R_{Lmin} = V_O/I_{Omax}$. Define a duty cycle range, that is, vary D from 0 to 1 in steps of 0.001.

3. Plot the lossy dc voltage transfer function M_{VDC} as a function of the duty cycle for $R_L = R_{Lmin}$ using the `plot` command.

4. Repeat the result for $R_L = R_{Lmax} = V_O/I_{Omin}$. Plot the curves of the voltage transfer function for the two different load resistances on the same figure window. Use `legend` or `text` commands to label the different curves clearly.

Post-lab Questions

1. What conclusions can be made from the figures obtained in Sections A, B, and C?
2. The efficiency of the buck converter decreases with increase in the load current. Justify.
3. Plot the curve of efficiency η versus duty cycle D using Equation (7.2) on MATLAB®. Replace the term M_{VDC} with D. Vary D from 0 to 1 in steps of 0.001 to obtain the curve.

8

Buck DC–DC Converter in DCM—Steady-State Simulation

Objectives

The objectives of this lab are:

- To design a pulse-width modulated buck dc–dc converter operating in discontinuous-conduction mode (DCM) for the specifications provided.
- To simulate and analyze the characteristics of the converter in steady state.
- To determine the losses and overall efficiency of the buck converter in DCM.

Specifications

The specifications of the buck converter are as provided in Table 8.1.

Pre-lab

For the above specifications, determine the values of all the components using the relevant design equations provided in Table B.1 in Appendix B. Also, determine the voltage and current stresses of the components. Make appropriate diode and MOSFET selections to match the desired specifications.

Quick Design

Choose:
$L = 2.4$ μH, ESR of the inductor $r_L = 0.05$ Ω, $R_{Lmin} = 1.2$ Ω, $D_{min} = 0.3162$, $D_{nom} = 0.378$, $D_{max} = 0.4714$, $C = 220$ μF, ESR of the capacitor $r_C = 0.025$ Ω.

Laboratory Manual for Pulse-Width Modulated DC–DC Power Converters, First Edition.
Marian K. Kazimierczuk and Agasthya Ayachit.
© 2016 John Wiley & Sons, Ltd. Published 2016 by John Wiley & Sons, Ltd.

Table 8.1 Parameters and their values

Parameter	Notation	Value
Minimum dc input voltage	V_{Imin}	24 V
Nominal dc input voltage	V_{Inom}	28 V
Maximum dc input voltage	V_{Imax}	32 V
DC output voltage	V_O	12 V
Switching frequency	f_s	100 kHz
Maximum output power	P_{Omax}	120 W
Minimum output power	P_{Omin}	0 W
Output voltage ripple	V_r	$< 0.06 V_O$

MOSFET: International Rectifier IRF150 n-channel power MOSFET with $V_{DSS} = 100$ V, $I_{SM} = 40$ A, $r_{DS} = 0.055\ \Omega$ at $T = 25°C$, $C_o = 100$ pF, and $V_t = 4$ V.

Diode: Fairchild Semiconductor MBR4040 with $V_{RRM} = 40$ V, $I_F = 40$ A, $R_F = 25$ mΩ, and $V_F = 0.4$ V.

Procedure

A. Simulation of the Buck Converter in DCM and its Analysis in Steady State

1. Construct the circuit of the buck converter shown in Figure 8.1 on the circuit simulator. Name all the nodes and components for convenience.
2. Let the input voltage V_I be equal to $V_{Inom} = 28$ V and the duty cycle is $D_{nom} = 0.378$. Also, set the value of the load resistance R_L to R_{Lmin}.
3. Enter the values of all the other components. Place a pulse voltage source in order to provide the gate-to-source voltage at the MOSFET gate and source terminals. Set time period = 10 μs, duty cycle/width = 0.378, and amplitude = 12 V. Let the rise time and fall time be equal to zero (optional).
4. Set simulation type to transient analysis. Set end time = 10 ms and a time step = 0.1 μs. Run the simulation.
5. Plot the following parameters after successful completion of the simulation. You may display the waveforms on different figure windows for better clarity.
   - Gate-to-source voltage v_{GS}, drain-to-source voltage v_{DS}, and diode voltage v_D.
   - Output voltage v_O, output current i_O, and output power p_O.
   - Inductor current i_L, diode current i_D, and MOSFET current i_S.

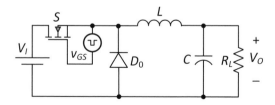

Figure 8.1 Circuit diagram of the PWM buck dc–dc converter.

Use the zoom option to display only the steady-state region.
6. Observe the inductor current waveform to ensure whether the current is in DCM. If the current is not in DCM, then decrease the value of the inductor and repeat the simulation.
7. For the above-mentioned waveforms, measure:
   - The average and peak-to-peak values of the current through the inductor L.
   - The maximum, intermediate, and average values of the voltage across the MOSFET S.
   - The minimum, intermediate, and average values of the voltage across the diode D_0.
   - The maximum and average values of the currents through the MOSFET and the diode.
   Ensure that the values obtained above match the desired specifications.
8. Repeat the steps above with a duty cycle and input voltage of $D_{max} = 0.4714$ and $V_{Imin} = 24$ V, respectively.
9. Repeat for a duty cycle $D_{min} = 0.3162$ and input voltage of $V_{Imax} = 32$ V.

B. Simulation of the Buck Converter to Determine the Power Losses and Overall Efficiency

1. Set up the converter to operate at nominal operating condition, that is, at $D = D_{nom}$ and $V_I = V_{Inom}$. Let $R_L = R_{Lmin}$ such that the converter delivers a maximum output power.
2. Set the simulation type to transient analysis and perform the simulation.
3. Plot the waveforms of the input power p_I and the output power p_O. Zoom in to the steady-state region.
4. Measure the average values of the input power and the output power. If the input power is negative, then consider only the magnitude of the average value of the input power.
5. Calculate the efficiency of the converter using $\eta = P_O/P_I$, where P_O is the average value of the output power and P_I is the average value of the input power, respectively.
6. This section may be repeated by plotting the power waveforms of all the components and then estimating their average values. All the power losses can be added to give the total power loss in the converter. Further, the efficiency can be estimated using $\eta = P_O/(P_{LS} + P_I)$, where P_O is the average value of the output power and P_{LS} is the sum of the average values of the power loss in individual components.
7. Repeat the above-mentioned activities in this section for $R_L = 10\,R_{Lmin}$ to determine the efficiency at $0.1P_{Omax}$.

Post-lab Question

Explain the operation of the buck converter in discontinuous-conduction mode. Draw the equivalent circuits of the converter representing its operation in different intervals of time.

9

Efficiency and DC Voltage Transfer Function of PWM Buck DC–DC Converter in DCM

Objectives

The objectives of this lab are:

- To design a PWM buck dc–dc converter in discontinuous-conduction mode (DCM) using the design equations.
- To analyze the variations in efficiency of the lossy buck converter in DCM at different load resistances and different input voltages.
- To observe the dependence of the lossy dc voltage transfer function on duty cycle.

Theory

The steady-state dc voltage transfer function of a converter is

$$M_{VDC} = \frac{V_O}{V_I}.$$

(9.1)

In terms of the circuit parameters, the transfer function under ideal conditions can be expressed as

$$M_{VDC} = \frac{2}{1 + \sqrt{1 + \dfrac{8f_s L I_O}{D^2 V_O}}},$$

(9.2)

Laboratory Manual for Pulse-Width Modulated DC–DC Power Converters, First Edition.
Marian K. Kazimierczuk and Agasthya Ayachit.
© 2016 John Wiley & Sons, Ltd. Published 2016 by John Wiley & Sons, Ltd.

resulting in

$$D = 1 - \frac{2f_s L I_O}{V_O}. \tag{9.3}$$

The overall converter efficiency is

$$\eta = \frac{1}{1 + \dfrac{P_{LS}}{P_O}}, \tag{9.4}$$

where P_{LS} is the total power loss in the converter. In terms of M_{VDC}, the total power loss is expressed as

$$P_{LS} = \left[\frac{2r_{DS}}{3} \sqrt{\frac{2M_{VDC}^2(1 - M_{VDC})}{f_s L R_L}} + \frac{2R_F}{3} \sqrt{\frac{2(1 - M_{VDC})^3}{f_s L R_L}} + \frac{2r_L}{3} \sqrt{\frac{2(1 - M_{VDC})}{f_s L R_L}} \right.$$

$$\left. + \frac{V_F(1 - M_{VDC})}{V_O} + \frac{f_s C_o R_L}{M_{VDC}^2} \right] P_O, \tag{9.5}$$

where r_L is the equivalent series resistance (ESR) of the buck inductor, r_{DS} is the on-state resistance of the MOSFET, D is the duty cycle, R_L is the load resistance, R_F is the forward resistance of the diode, r_C is the ESR of the capacitor, V_F is the forward voltage drop of the diode, V_O is the output voltage, f_s is the switching frequency, and C_o is the output capacitance of the MOSFET. In terms of the duty cycle, the total power loss P_{LS} can be expressed as

$$P_{LS} = \left[\frac{2r_{DS}}{3} \sqrt{\frac{2D^2(1 - D)}{f_s L R_L}} + \frac{2R_F}{3} \sqrt{\frac{2(1 - D)^3}{f_s L R_L}} + \frac{2r_L}{3} \sqrt{\frac{2(1 - D)}{f_s L R_L}} + \frac{V_F(1 - D)}{V_O} + \frac{f_s C_o R_L}{D^2} \right] P_O. \tag{9.6}$$

The steady-state dc voltage transfer function of the lossy buck converter in DCM is

$$M_{VDC} = \frac{2}{1 + \sqrt{1 + \dfrac{8f_s L I_O}{\eta D^2 V_O}}} = \frac{2}{1 + \sqrt{1 + \dfrac{8f_s L}{\eta D^2 R_L}}}. \tag{9.7}$$

Specifications

The specifications of the buck converter are as provided in Table 9.1.

Pre-lab

For the above specifications, find the values of all the components and specifications for the buck dc–dc converter operating in DCM using the relevant design equations given in Table B.1 in Appendix B.

Table 9.1   Parameters and component values

Parameters	Notation	Value
Minimum dc input voltage	V_{Imin}	24 V
Maximum dc input voltage	V_{Imax}	28 V
Maximum dc input voltage	V_{Imax}	32 V
DC output voltage	V_O	12 V
Switching frequency	f_s	100 kHz
Maximum output power	P_{Omax}	120 W
Minimum output power	P_{Omin}	0 W
Output voltage ripple	V_r	$< 0.06V_O$

Quick Design

Choose:

$L = 2.4$ μH, ESR of the inductor $r_L = 0.05$ Ω, $R_{Lmin} = 1.2$ Ω, $D_{min} = 0.3162$, $D_{nom} = 0.378$, $D_{max} = 0.4714$, $C = 220$ μF, ESR of the capacitor $r_C = 0.025$ Ω.

MOSFET: International Rectifier IRF150 n-channel power MOSFET with $V_{DSS} = 100$ V, $I_{SM} = 40$ A, $r_{DS} = 0.055$ Ω at $T = 25°C$, $C_o = 100$ pF, and $V_t = 4$ V.

Diode: Fairchild Semiconductor MBR4040 with $V_{RRM} = 40$ V, $I_F = 40$ A, $R_F = 25$ mΩ, and $V_F = 0.4$ V.

Procedure

A. Efficiency of the Buck Converter as a Function of the Input Voltage at Different Load Conditions

1. The expressions for the efficiency η of the lossy buck converter and the total power loss P_{LS} are as given in Equations (9.4) and (9.5), respectively. Replace the term M_{VDC} in Equation (9.5) with V_O/V_I. Initially, let $R_L = 1.2$ Ω.
2. Define the equation for the efficiency as given in Equation (9.4) and (9.5) on MATLAB® along with all the specifications and their values. Define an input voltage range, that is, vary V_I from 24 to 32 V in steps of 0.01 V.
3. Plot the efficiency as a function of the input voltage for $R_L = 1.2$ Ω.
4. Repeat the activity for $R_L = 2.4$ Ω and $R_L = 12$ Ω. Plot all the three curves on the same figure window. Use `legend` or `text` commands to label the different curves clearly.

B. Efficiency of the Buck Converter as a Function of the Output Current at Minimum, Nominal, and Maximum Input Voltages

1. Use the code developed in Section A. In this section, fix the value of the input voltage V_I at V_{Imin}.
2. Define the load resistance variable as $R_L = V_O/I_O$. Let the output current I_O vary from 0 to 10 A in steps of 0.01 A.

3. Plot the efficiency as a function of the output current for $V_I = V_{Imin}$ using the `plot` command.

4. Repeat the above steps by changing the input voltage to V_{Inom} and then to V_{Imax}. Plot the efficiency curves for the three different input voltages on the same figure window. Use `legend` or `text` commands to label the different curves clearly.

C. DC Voltage Transfer Function of the Buck Converter as a Function of the Duty Cycle

1. The equation for the total power loss P_{LS} in the boost converter in terms of duty cycle D is given in Equation (9.6). Redefine the equation for efficiency as given in Equation (9.4) using Equation (9.6). Initially, let $R_L = R_{Lmin} = 1.2\ \Omega$.

2. Next, define the equation for the dc voltage transfer function of the lossy buck converter as given in Equation (9.7). Let the range of the duty cycle D be from 0 to 1 and increased in steps of 0.001.

3. Plot the lossy dc voltage transfer function M_{VDC} as a function of the duty cycle D for $R_L = R_{Lmin}$ using the `plot` command.

4. Repeat the activity for $R_L = 2.4\ \Omega$ and $R_L = 12\ \Omega$. Plot the curves of the voltage transfer function for the three different load resistances on the same figure window. Use `legend` or `text` commands to label the different curves clearly.

Post-lab Question

Consider Eq. 9.3. Replace the ratio I_O/V_O by R_L. The duty cycle D depends on the load resistance R_L. Plot the duty cycle as a function of the load resistance for three different values of the input voltage. Vary R_L from 1 Ω to 12 Ω. Set the input voltage at $V_I = V_{Imin}$. Using the `plot` command, plot the curve D vs. R_L. Repeat the activity for $V_I = V_{Inom}$ and $V_I = V_{Imax}$. Present all the plots on the same figure window. Use `legend` or `text` commands to label the different curves clearly.

10

High-Side Gate-Drive Circuit for Buck DC–DC Converter

Objective

The objective of this lab is to simulate a buck converter driven by a high-side gate-drive circuit.

Pre-lab

Obtain the values of the components in the buck dc–dc converter from Lab 6. You may consider redesigning the converter for the specifications provided for completeness.

Quick Design

Assume an overall efficiency of $\eta = 85\%$. Choose:

$L = 40$ μH, ESR of the inductor $r_L = 0.05$ Ω, $R_{Lmin} = 1.2$ Ω, $R_{Lmax} = 12$ Ω, $D_{min} = 0.441$, $D_{nom} = 0.506$, $D_{max} = 0.588$, $C = 100$ μF, ESR of the capacitor $r_C = 50$ mΩ.

MOSFET: International Rectifier IRF150 n-channel power MOSFET with $V_{DSS} = 100$ V, $I_{SM} = 40$ A, $r_{DS} = 55$ mΩ at $T = 25°C$, $C_o = 100$ pF, and $V_t = 4$ V.

Diode: ON Semiconductor MUR1060 with $V_{RRM} = 60$ V, $I_F = 20$ A, $R_F = 25$ mΩ, and $V_F = 0.4$ V.

Gate-drive circuit: $V_{CC} = 15$ V, $C_1 = 1$ μF, $R_2 = 52$ kΩ, $R_3 = 10$ kΩ, $R_1 = 50$ Ω, $C_B = 0.1$ μF. Use 1N4148 for diode D_B, Q2N2222 for the BJT Q_1, and Q2N2907 for the BJT Q_2. M can be an ideal MOSFET or any small-signal MOSFET such as Q2N7000.

Procedure

1. Construct the circuit of the buck converter with the gate-drive circuit shown in Figure 10.1 on the circuit simulator. Name all the nodes and components for convenience.

Laboratory Manual for Pulse-Width Modulated DC–DC Power Converters, First Edition.
Marian K. Kazimierczuk and Agasthya Ayachit.
© 2016 John Wiley & Sons, Ltd. Published 2016 by John Wiley & Sons, Ltd.

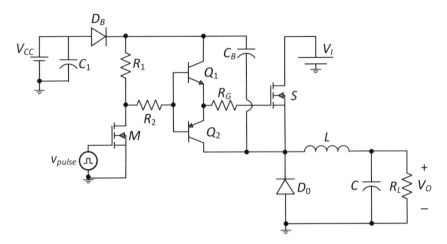

Figure 10.1 Circuit diagram of the PWM buck dc–dc converter.

2. Initially, simulate the converter at $V_I = V_{Inom} = 28$ V, $R_L = R_{Lmin} = 1.2\,\Omega$. Set the duty cycle of the MOSFET M to be equal to $D = D_{nom} = 0.506$. Connect a pulse voltage source between the gate and source terminals of the MOSFET M. Set `time period = 10 μs`, `duty cycle/width = 0.506`, and `amplitude = 12` V. Let the `rise time` and `fall time` be equal to zero (optional).

3. Set simulation type to `transient analysis`. Set `end time = 10 ms` and a `time step = 0.1 μs`. Run the simulation.

4. Plot the following parameters after successful completion of the simulation. You may display the waveforms on different figure windows for better clarity.
   - Gate-to-source voltage v_{GS}, drain-to-source voltage v_{DS}, and diode voltage v_D.
   - Output voltage v_O, output current i_O, and output power p_O.
   - Inductor current i_L, diode current i_D, and MOSFET current i_S.
   Zoom in to display only the steady-state region.

5. Observe the inductor current waveform to ensure whether the current is in CCM. If the current is not in CCM, then increase the value of the inductor and repeat the simulation.

6. For the above-mentioned waveforms, measure:
   - The average and peak-to-peak values of the current through the inductor L.
   - The maximum and average values of the voltage across the MOSFET S.
   - The minimum and average values of the voltage across the diode D_0.
   - The maximum and average values of the currents through the MOSFET and the diode.
   Ensure that the values obtained above match the desired specifications.

Post-lab Questions

1. Explain the need for the high-side gate-drive circuit for the buck converter.
2. What are the differences between high-side and low-side gate-drive schemes?
3. Explain the necessity for the bootstrap diode and bootstrap capacitance used in the circuit.
4. What are the characteristics of a good gate-drive circuit?

11

Quadratic Buck DC–DC Converter in CCM—Steady-State Simulation

Objective

The objective of this lab is to simulate the pulse-width modulated (PWM) quadratic buck dc–dc converter operating in continuous-conduction mode (CCM).

Specifications

The specifications of the quadratic buck converter are as provided in Table 11.1.

Quick Design

Assume an overall efficiency of $\eta = 100\%$. Choose:

$L_1 = 40\ \mu H$, $L_1 = 27\ \mu H$, $C_1 = 16\ \mu F$, $C_2 = 18\ \mu F$, $R_L = 11.1\ \Omega$, $D = 0.64$.

You may consider using the ideal MOSFET and diodes for this experiment. Alternatively, you may choose:

MOSFET: International Rectifier IRF150 n-channel power MOSFET with $V_{DSS} = 100$ V, $I_{SM} = 40$ A, $r_{DS} = 55$ mΩ at $T = 25°C$, $C_o = 100$ pF, and $V_t = 4$ V.

Diode: ON Semiconductor MUR1060 with $V_{RRM} = 60$ V, $I_F = 20$ A, $R_F = 25$ mΩ, and $V_F = 0.4$ V.

Procedure

A. Simulation of the Quadratic Buck Converter in CCM

1. Construct the circuit of the quadratic buck converter shown in Figure 11.1 on the circuit simulator. Name all the nodes and components for convenience.
2. Assign the values of all the components. Connect a pulse voltage source in order to provide the gate-to-source voltage at the MOSFET gate and source terminals. Set `time`

Laboratory Manual for Pulse-Width Modulated DC–DC Power Converters, First Edition.
Marian K. Kazimierczuk and Agasthya Ayachit.
© 2016 John Wiley & Sons, Ltd. Published 2016 by John Wiley & Sons, Ltd.

Table 11.1 Parameters and their values

Parameter	Notation	Value
Nominal dc input voltage	V_{Inom}	24 V
DC output voltage	V_O	10 V
Switching frequency	f_s	110 kHz
Rated output power	P_O	10 W
Output voltage ripple	V_r	$< 0.01 V_O$

period = 9.09 µs, duty cycle/width = 0.581, and amplitude = 12 V. Let the rise time and fall time be equal to zero (optional).

3. Set simulation type to transient analysis. Set end time = 10 ms and a time step = 0.1 µs. Run the simulation.
4. Plot the following parameters after successful completion of the simulation. You may display the waveforms on different figure windows for better clarity.
   - Gate-to-source voltage v_{GS} and drain-to-source voltage v_{DS}.
   - Diode voltages v_{D0}, v_{D1}, and v_{D2}.
   - Output voltage v_O, output current i_O, and output power p_O.
   - Inductor currents i_{L1}, and i_{L2}.
   - Diode currents i_{D0}, i_{D1}, and i_{D2}, and MOSFET current i_S.
   Use the zoom option to display only the steady-state region.
5. Observe the inductor current waveforms to ensure whether the current is in CCM. If the current is not in CCM, then increase the values of the both inductors and repeat the simulation.
6. For the above-mentioned waveforms, measure:
   - The average and peak-to-peak values of the currents through the inductors.
   - The maximum and average values of the voltage across the MOSFET.
   - The minimum and average values of the voltages across the diodes.
   - The maximum and average values of the currents through the MOSFET and the diodes.

Post-lab Questions

1. Represent the values obtained above in the form of a well-organized table.
2. Explain the operation of the quadratic buck converter in CCM.
3. Identify the basic differences between the basic buck and quadratic buck converters in CCM.

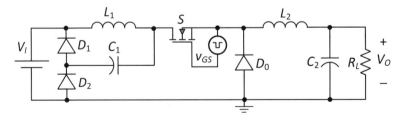

Figure 11.1 Circuit diagram of the PWM quadratic buck dc–dc converter.

4. Derive the equation for the dc voltage transfer function of the quadratic buck converter in CCM.
5. What are the advantages of using converters with wide duty ratio?
6. The dc voltage transfer function of the quadratic buck converter is $M_{VDC} = D^2$. The dc voltage transfer function of the conventional buck converter is $M_{VDC} = D$. Plot the two dc voltage transfer functions as a function of duty cycle on MATLAB®. Compare the results and comment on the differences between the nature of the two curves.

12

Buck–Boost DC–DC Converter in CCM—Steady-State Simulation

Objectives

The objectives of this lab are:

- To design a pulse-width modulated (PWM) buck–boost dc–dc converter operating in continuous-conduction mode (CCM) for the design specifications provided.
- To simulate the converter and analyze its characteristics in steady state.
- To estimate the overall efficiency of the buck–boost converter.

Specifications

The specifications of the buck–boost converter are as provided in Table 12.1.

Pre-lab

For the specifications provided, find the values of all the components and parameters for the buck–boost dc–dc converter operating in CCM using the relevant design equations provided in Table A.1 in Appendix A.

Quick Design

Assume an overall efficiency of $\eta = 85\%$. Choose:

$L = 30$ µH, ESR of the inductor $r_L = 0.05$ Ω, $R_{Lmin} = 1.2$ Ω, $R_{Lmax} = 12$ Ω, $D_{min} = 0.306$, $D_{nom} = 0.335$, $D_{max} = 0.370$, $C = 2.2$ mF, ESR of the capacitor $r_C = 6$ mΩ.

MOSFET: International Rectifier IRF142 n-channel power MOSFET with $V_{DSS} = 100$ V, $I_{SM} = 24$ A, $r_{DS} = 0.11$ mΩ at $T = 25°C$, $C_o = 100$ pF, and $V_t = 4$ V.

Laboratory Manual for Pulse-Width Modulated DC–DC Power Converters, First Edition.
Marian K. Kazimierczuk and Agasthya Ayachit.
© 2016 John Wiley & Sons, Ltd. Published 2016 by John Wiley & Sons, Ltd.

Table 12.1 Parameters and their values

Parameter	Notation	Value
Minimum dc input voltage	V_{Imin}	24 V
Nominal dc input voltage	V_{Inom}	28 V
Maximum dc input voltage	V_{Imax}	32 V
DC output voltage	V_O	12 V
Switching frequency	f_s	100 kHz
Maximum output current	I_{Omax}	10 A
Minimum output current	I_{Omin}	1 A
Output voltage ripple	V_r	$< 0.01V_O$

Diode: ON Semiconductor MUR2510 with $V_{RRM} = 100$ V, $I_F = 25$ A, $R_F = 20$ mΩ, and $V_F = 0.7$ V.

Procedure

A. Simulation and Analysis of the Buck–Boost Converter in Steady State

1. Construct the circuit of the buck–boost converter shown in Figure 12.1 on the circuit simulator. Name all the nodes and components for convenience.
2. Enter the values of all the components. Initially, let the input voltage V_I be equal to $V_{Imax} = 32$ V and the duty cycle be $D_{min} = 0.306$. Set the value of the load resistance to $R_{Lmin} = 1.2$ Ω.
3. Enter the values of all the components. Connect a pulse voltage source between the gate and source terminals of the MOSFET. Set `time period` $= 10$ μs, `duty cycle/width` $= 0.335$, and `amplitude` $= 12$ V. Let the `rise time` and `fall time` be equal to zero (optional).
4. Set simulation type to `transient analysis`. Set `end time` $= 10$ ms and a `time step` $= 0.1$ μs. Run the simulation.
5. Plot the following parameters after successful completion of the simulation. You may display the waveforms on different figure windows for better clarity.
   - Gate-to-source voltage v_{GS}, drain-to-source voltage v_{DS}, and diode voltage v_D.
   - Output voltage V_O, output current I_O, and output power P_O.
   - Inductor current i_L, diode current i_D, and MOSFET current i_S.

   Zoom in to display only the steady-state region.

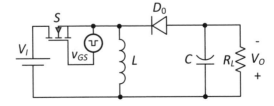

Figure 12.1 Circuit diagram of the PWM buck–boost dc–dc converter.

6. Observe the inductor current waveform to ensure whether the current is in CCM. If the current is not in CCM, then increase the value of the inductor and repeat the simulation.
7. For the above-mentioned waveforms, measure:
   - The average and peak-to-peak values of the current through the inductor L.
   - The maximum and average values of the voltage across the MOSFET S.
   - The minimum and average values of the voltage across the diode D_0.
   - The maximum and average values of the currents through the MOSFET and the diode.
   Ensure that the values obtained above match the desired specifications.
8. Repeat the steps above with a duty cycle of $D_{nom} = 0.335$ and input voltage $V_{Inom} = 28$ V. Also, repeat for duty cycle $D_{max} = 0.370$ and an input voltage of $V_{Imin} = 24$ V.

B. Estimation of the Overall Efficiency of the Buck–Boost Converter

1. Set up the converter to operate at nominal operating condition, that is, at $D = D_{nom}$ and $V_I = V_{Inom}$. Let $R_L = R_{Lmin}$ such that the converter delivers a maximum output power.
2. Set the simulation type to transient analysis and perform the simulation.
3. Plot the waveforms of the input power p_I and the output power p_O. Zoom in to the steady-state region.
4. Measure the average values of the input power and the output power. If the input power is negative, then consider only the magnitude of the average value of the input power.
5. Calculate the efficiency of the converter using $\eta = P_O/P_I$, where P_O is the average value of the output power and P_I is the average value of the input power, respectively.
6. This section may be repeated by plotting the power waveforms of all the components and then estimating their average values. All the power losses can be added to give the total power loss in the converter. Further, the efficiency can be estimated using $\eta = P_O/(P_{LS} + P_I)$, where P_O is the average value of the output power and P_{LS} is the sum of the average values of the power loss in individual components.

Post-lab Questions

1. Tabulate all the values obtained in Section A.
2. Explain the operation of the buck–boost dc–dc converter in CCM.
3. When the polarity of the diode is reversed in the buck–boost converter, explain why the circuit obtained cannot function as a dc–dc converter.
4. Draw the waveform of the inductor current. Show the variation in the slope of the current for the three different values of the input voltage.
5. In Section B, identify the component which has the highest power loss.

13

Efficiency and DC Voltage Transfer Function of PWM Buck–Boost DC–DC Converter in CCM

Objectives

The objectives of this lab are:

- To design a PWM buck–boost dc–dc converter in CCM for the given design specifications.
- To analyze the variations in efficiency of the lossy buck–boost converter for different load resistances and different input voltages.
- To analyze the variation in the lossy dc voltage transfer function with a change in duty cycle.

Theory

The steady-state dc voltage transfer function of a converter is

$$M_{VDC} = \frac{V_O}{V_I}. \tag{13.1}$$

The overall converter efficiency is

$$\eta = \frac{N_\eta}{D_\eta}, \tag{13.2}$$

where

$$N_\eta = 1 - M_{VDC}\left(\frac{2r_L + r_{DS} + R_F + r_C}{R_L}\right) + \left\{ \left[1 - M_{VDC}\left(\frac{2r_L + r_{DS} + R_F + r_C}{R_L}\right)\right]^2 \right.$$
$$\left. - \frac{4M_{VDC}^2(r_L + r_{DS})}{R_L}\left[1 + \frac{R_F + r_L}{R_L} + \frac{V_F}{V_O} + \frac{f_s C_o R_L (1 + M_{VDC})^2}{M_{VDC}^2}\right] \right\}^{\frac{1}{2}},$$

Laboratory Manual for Pulse-Width Modulated DC–DC Power Converters, First Edition.
Marian K. Kazimierczuk and Agasthya Ayachit.
© 2016 John Wiley & Sons, Ltd. Published 2016 by John Wiley & Sons, Ltd.

and

$$D_\eta = 2 \left[1 + \frac{R_F + r_L}{R_L} + \frac{V_F}{V_O} + \frac{f_s C_o R_L (1 + M_{VDC})^2}{M_{VDC}^2} \right],$$

where r_L is the equivalent series resistance (ESR) of the inductor, r_{DS} is the on-state resistance of the MOSFET, D is the duty cycle, R_L is the load resistance, R_F is the forward resistance of the diode, r_C is the ESR of the capacitor, V_F is the forward voltage drop of the diode, V_O is the output voltage, f_s is the switching frequency, and C_o is the output capacitance of the MOSFET.

The steady-state dc voltage transfer function of the lossy buck–boost converter is

$$M_{VDC} = \frac{\eta D}{1 - D}. \tag{13.5}$$

Specifications

The specifications of the buck–boost converter are as provided in Table 13.1.

Pre-lab

For the above specifications, find the values of all the components and parameters of the buck–boost dc–dc converter operating in CCM using the relevant design equations provided in Table A.1 in Appendix A.

Quick Design

Assume an overall efficiency of $\eta = 85\%$. Choose:

$L = 30$ μH, ESR of the inductor $r_L = 0.05$ Ω, $R_{Lmin} = 1.2$ Ω, $R_{Lmax} = 12$ Ω, $D_{min} = 0.306$, $D_{nom} = 0.335$, $D_{max} = 0.370$, $C = 2.2$ mF, ESR of the capacitor $r_C = 6$ mΩ.

MOSFET: International Rectifier IRF142 n-channel power MOSFET with $V_{DSS} = 100$ V, $I_{SM} = 24$ A, $r_{DS} = 0.11$ mΩ at $T = 25°C$, $C_o = 100$ pF, and $V_t = 4$ V.

Table 13.1 Parameters and their values

Parameter	Notation	Value
Minimum dc input voltage	V_{Imin}	24 V
Nominal dc input voltage	V_{Inom}	28 V
Maximum dc input voltage	V_{Imax}	32 V
DC output voltage	V_O	12 V
Switching frequency	f_s	100 kHz
Maximum output current	I_{Omax}	10 A
Minimum output current	I_{Omin}	1 A
Output voltage ripple	V_r	$< 0.01 V_O$

Diode: ON Semiconductor MUR2510 with $V_{RRM} = 100$ V, $I_F = 25$ A, $R_F = 20$ mΩ, and $V_F = 0.7$ V.

Procedure

A. Efficiency of the Buck–Boost Converter as a Function of the Input Voltage at Full and Light Load Conditions

1. The equation for the efficiency of the lossy buck–boost converter is as given in Equation (13.2). Replace the term M_{VDC} in Equation (13.2) with the term V_O/V_I. Initially, let $R_L = R_{Lmin} = V_O/I_{Omax}$.
2. Define the equation for the efficiency on MATLAB® along with all the specifications and their values. Define an input voltage range, that is, vary V_I from 20 to 40 V in steps of 0.01 V.
3. Plot the efficiency as a function of the input voltage for $R_L = R_{Lmin}$.
4. Repeat the activity by replacing R_L with $R_{Lmax} = V_O/I_{Omin}$. Plot the two curves on the same figure window. Use `legend` or `text` commands to label the different curves clearly.

B. Efficiency of the Buck–Boost Converter as a Function of the Output Current at Minimum, Nominal, and Maximum Input Voltages

1. Use the code developed in the previous section. Assign a fixed value of input voltage. Let $V_I = V_{Imin}$.
2. Replace the term R_L with V_O/I_O. Define a range for the output current, that is, vary I_O from 0 to 10 A in steps of 0.001 A.
3. Plot the curve of efficiency as a function of the output current at $V_I = V_{Imin}$.
4. Next, repeat the activity by replacing the input voltage with V_{Inom} and then with V_{Imax}. The three curves must be plotted on the same figure window. Use `legend` or `text` commands to label the different curves clearly.

C. DC Voltage Transfer Function of the Buck–Boost Converter as a Function of the Duty Cycle

1. Use the code for the efficiency as developed in Section A. Replace the term M_{VDC} in Equation (13.2) with $D/(1 - D)$.
2. Define the equation for the dc voltage transfer function of the lossy buck–boost converter as given in Equation (13.5). Initially, let $R_L = R_{Lmin} = V_O/I_{Omax}$. Define a duty cycle range, that is, vary D from 0 to 1 in steps of 0.001.
3. Plot the dc voltage transfer function M_{VDC} as a function of the duty cycle for $R_L = R_{Lmin}$ using the `plot` command.
4. Repeat the activity for $R_L = R_{Lmax} = V_O/I_{Omin}$. Plot the curves of the voltage transfer function for the three different load resistances on the same figure window. Use `legend` or `text` commands to label the different curves clearly.

Post-lab Question

Consider Eq. 13.2. Replace the term M_{VDC} by D. Plot the efficiency as a function of the duty cycle for two different values of the load resistance. Vary D from 0 to 1 in steps of 0.001. Set the load resistance at $R_L = R_{Lmin}$. Using the plot command, plot the curve η vs. D. Repeat the activity for $R_L = R_{Lmax}$. Present all the plots on the same figure window. Use legend or text commands to label the different curves clearly.

14

Buck–Boost DC–DC Converter in DCM—Steady-State Simulation

Objectives

The objectives of this lab are:

- To design a pulse-width modulated (PWM) buck–boost dc–dc converter operating in discontinuous-conduction mode (DCM) for the design specifications provided.
- To simulate the converter and analyze its characteristics in steady state.
- To estimate the overall efficiency of the buck–boost converter in DCM.

Specifications

The specifications of the buck–boost converter are as provided in Table 14.1.

Pre-lab

For the above specifications, find the values of all the components and parameters for the buck–boost dc–dc converter operating in DCM using the relevant design equations provided in Table B.1 in Appendix B.

Quick Design

Choose:

$L = 2.2$ μH, ESR of the inductor $r_L = 0.01$ Ω, $R_{Lmin} = 1.2$ Ω, $D_{min} = 0.306$, $D_{nom} = 0.335$, $D_{max} = 0.370$, $C = 1.8$ mF, ESR of the capacitor $r_C = 0.025$ Ω.

MOSFET: International Rectifier IRF142 n-channel power MOSFET with $V_{DSS} = 100$ V, $I_{SM} = 24$ A, $r_{DS} = 0.11$ Ω at $T = 25°C$, $C_o = 100$ pF, and $V_t = 4$ V.

Laboratory Manual for Pulse-Width Modulated DC–DC Power Converters, First Edition.
Marian K. Kazimierczuk and Agasthya Ayachit.
© 2016 John Wiley & Sons, Ltd. Published 2016 by John Wiley & Sons, Ltd.

Table 14.1   Parameters and their values

Parameter	Notation	Value
Minimum dc input voltage	V_{Imin}	24 V
Nominal dc input voltage	V_{Inom}	28 V
Maximum dc input voltage	V_{Imax}	32 V
DC output voltage	V_O	12 V
Switching frequency	f_s	100 kHz
Maximum output current	I_{Omax}	10 A
Minimum output power	O_{Omin}	0 A
Output voltage ripple	V_r	$< 0.06V_O$

Diode: Fairchild Semiconductor MUR2510 with $V_{RRM} = 100$ V, $I_F = 25$ A, $R_F = 20$ mΩ, and $V_F = 0.7$ V.

Procedure

A. *Simulation and Analysis of the Buck–Boost Converter in DCM*

1. Construct the circuit of the buck–boost converter shown in Figure 14.1 on the circuit simulator. Name all the nodes and components.
2. Enter the values of all the components. Initially, let the input voltage V_I be equal to $V_{Imax} = 32$ V and the duty cycle be $D_{min} = 0.306$. Set the value of the load resistance to $R_{Lmin} = 1.2$ Ω.
3. Place a pulse voltage source in order to provide the gate-to-source voltage at the MOSFET terminals. Set `time period = 10` µs, `duty cycle/width = 0.306`, and `ampli-tude = 12` V. Let the `rise time` and `fall time` be equal to zero (optional).
4. Set simulation type to `transient analysis`. Set `end time = 10 ms` and a `time step = 0.1` µs. Run the simulation.
5. Plot the following parameters after successful completion of the simulation. You may display the waveforms on different figure windows for better clarity.
   - Gate-to-source voltage v_{GS}, drain-to-source voltage v_{DS}, and diode voltage v_D.
   - Output voltage v_O, output current i_O, and output power p_O.
   - Inductor current i_L, diode current i_D, and MOSFET current i_S.
   Use the `zoom` option to display only on the steady-state region.
6. Observe the inductor current waveform to ensure whether the current is in DCM. If the current is not in DCM, then reduce the value of the inductor and repeat the simulation.

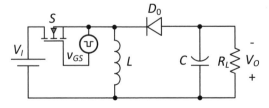

Figure 14.1   Circuit diagram of the PWM buck–boost dc–dc converter.

7. For the above-mentioned waveforms, measure:
   - The average and peak-to-peak values of the current through the inductor L.
   - The maximum, intermediate, and average values of the voltage across the MOSFET S.
   - The minimum, intermediate, and average values of the voltage across the diode D_0.
   - The maximum, intermediate, and average values of the currents through the MOSFET and the diode.

   Ensure that the values obtained above match the desired specifications.
8. Repeat the steps above with duty cycle $D_{max} = 0.370$ and input voltage $V_{Imin} = 24$ V.
9. Next, repeat with duty cycle $D_{nom} = 0.335$ and input voltage of $V_{Inom} = 28$ V.

B. Estimation of the Overall Efficiency of the Buck–Boost Converter

1. Set up the converter to operate at nominal operating condition, that is, at $D = D_{nom}$ and $V_I = V_{Inom}$. Let $R_L = R_{Lmin}$ such that the converter is operating at maximum output power.
2. Set the simulation type to transient analysis and perform the simulation.
3. Plot the waveforms of the input power p_I and the output power p_O. Zoom in to the steady-state region.
4. Measure the average values of the input power and the output power. If the input power is negative, then consider only the magnitude of the average value of the input power.
5. Calculate the efficiency of the converter using $\eta = P_O/P_I$, where P_O is the average value of the output power and P_I is the average value of the input power, respectively.
6. This section may be repeated by plotting the power waveforms of all the components and then estimating their average values. All the power losses can be added to give the total power loss in the converter. Further, the efficiency can be estimated using $\eta = P_O/(P_{LS} + P_I)$, where P_O is the average value of the output power and P_{LS} is the sum of the average values of the power loss in individual components.

Post-lab Questions

1. What is the range of the dc voltage transfer function for the lossless and lossy buck–boost converter?
2. Is the current through the filter capacitor continuous in the buck–boost converter? What is the peak-to-peak value of the capacitor current?
3. How does the efficiency of the buck–boost converter change with the duty cycle?

15

Efficiency and DC Voltage Transfer Function of PWM Buck–Boost DC–DC Converter in DCM

Objectives

The objectives of this lab are:

- To design a PWM buck–boost dc–dc converter in DCM using the design equations.
- To analyze the variations in efficiency of the lossy buck–boost converter in DCM at different loads and different input voltages.
- To analyze the variation in lossy dc voltage transfer function for changes in the duty cycle.

Theory

The steady-state dc voltage transfer function of a converter is

$$M_{VDC} = \frac{V_O}{V_I}. \tag{15.1}$$

In terms of the circuit parameters, the transfer function under ideal conditions can be expressed as

$$M_{VDC} = D\sqrt{\frac{V_O}{2f_s L I_O}}, \tag{15.2}$$

resulting in

$$D = 1 - \sqrt{\frac{2f_s L I_O}{V_O}}. \tag{15.3}$$

Laboratory Manual for Pulse-Width Modulated DC–DC Power Converters, First Edition.
Marian K. Kazimierczuk and Agasthya Ayachit.
© 2016 John Wiley & Sons, Ltd. Published 2016 by John Wiley & Sons, Ltd.

The overall converter efficiency is

$$\eta = \frac{1}{1 + \frac{P_{LS}}{P_O}}, \tag{15.4}$$

where P_{LS} is the total power loss in the converter in terms of M_{VDC} is

$$P_{LS} = \left[\frac{2r_{DS}}{3} M_{VDC} \sqrt{\frac{2}{f_s L R_L}} + \frac{2R_F}{3} \sqrt{\frac{2}{f_s L R_L}} + \frac{2r_L}{3} M_{VDC} \sqrt{\frac{2}{f_s L R_L}} \left(1 + \frac{1}{M_{VDC}} \right) \right.$$
$$\left. + \frac{V_F}{V_O} + f_s C_o R_L \left(\frac{1}{M_{VDC}} + 1 \right)^2 \right] P_O, \tag{15.5}$$

where r_L is the equivalent series resistance (ESR) of the buck inductor, r_{DS} is the on-state resistance of the MOSFET, D is the duty cycle, R_L is the load resistance, R_F is the forward resistance of the diode, r_C is the ESR of the capacitor, V_F is the forward voltage drop of the diode, V_O is the output voltage, f_s is the switching frequency, and C_o is the output capacitance of the MOSFET. In terms of the duty cycle, the total power loss P_{LS} can be expressed as

$$P_{LS} = \left[\frac{2Dr_{DS}}{3f_s L} + \frac{2R_F}{3} \sqrt{\frac{2}{f_s L R_L}} + \frac{2Dr_L}{3f_s L} \left(1 + \sqrt{\frac{2f_s L}{D^2 R_L}} \right) \right.$$
$$\left. + \frac{V_F}{V_O} + f_s C_o R_L \left(\sqrt{\frac{2f_s L}{D^2 R_L}} + 1 \right)^2 \right] P_O. \tag{15.6}$$

The steady-state dc voltage transfer function of the lossy buck–boost converter in DCM is

$$M_{VDC} = D \sqrt{\frac{\eta R_L}{2f_s L}} = D \sqrt{\frac{\eta V_O}{2f_s L I_O}}. \tag{15.7}$$

Specifications

The specifications of the buck–boost converter are provided in Table 15.1.

Pre-lab

For the above specifications, find the values of all the components and parameters for the buck–boost dc–dc converter operating in DCM using the relevant design equations provided in Table B.1 in Appendix B.

Table 15.1 Parameters and their values

Parameter	Notation	Value
Minimum dc input voltage	V_{Imin}	24 V
Nominal dc input voltage	V_{Inom}	28 V
Maximum dc input voltage	V_{Imax}	32 V
DC output voltage	V_O	12 V
Switching frequency	f_s	100 kHz
Maximum output current	I_{Omax}	10 A
Minimum output power	O_{Omin}	0 A
Output voltage ripple	V_r	$< 0.06 V_O$

Quick Design

Choose:

$L = 2.2 \ \mu H$, ESR of the inductor $r_L = 0.01 \ \Omega$, $R_{Lmin} = 1.2 \ \Omega$, $D_{min} = 0.306$, $D_{nom} = 0.335$, $D_{max} = 0.370$, $C = 1.8$ mF, ESR of the capacitor $r_C = 0.025 \ \Omega$.

MOSFET: International Rectifier IRF142 n-channel power MOSFET with $V_{DSS} = 100$ V, $I_{SM} = 24$ A, $r_{DS} = 0.11 \ \Omega$ at $T = 25°C$, $C_o = 100$ pF, and $V_t = 4$ V.

Diode: Fairchild Semiconductor MUR2510 with $V_{RRM} = 100$ V, $I_F = 25$ A, $R_F = 20$ mΩ, and $V_F = 0.7$ V.

Procedure

A. Efficiency of the Buck–Boost Converter as a Function of the Input Voltage at Different Load Conditions

1. The equation for the efficiency of the lossy buck–boost converter is given in Equation (15.4). Replace the term M_{VDC} in Equation (15.5) with V_O/V_I. Initially, let $R_L = 12 \ \Omega$.
2. Define the equation for the efficiency as given in Equations (15.4) and (15.5) on MATLAB® along with all the specifications and their values. Define an input voltage range, that is, vary V_I from 24 to 32 V in steps of 0.01 V.
3. Plot the efficiency as a function of the input voltage for $R_L = 12 \ \Omega$.
4. Repeat the activity for $R_L = 2.4 \ \Omega$ and $R_L = 1.2 \ \Omega$. Plot all the three curves on the same figure window. Use legend or text commands to label the different curves clearly.

B. Efficiency of the Buck–Boost Converter as a Function of the Output Current at Minimum, Nominal, and Maximum Input Voltages

1. The equation for the efficiency of the lossy buck–boost converter is given in Equation (15.4). Replace the term M_{VDC} in Equation (15.5) with V_O/V_I. Also, replace the term R_L with V_O/I_O. Initially, let $V_I = V_{Imin} = 24$ V.
2. Define the equation for the efficiency on MATLAB® along with all the specifications and their values. Define an output current range, that is, vary I_O from 0 to 10 A in steps of 0.001 A.

3. Plot the efficiency as a function of the output current for $V_I = V_{Imin}$.
4. Repeat the activity for $V_{Inom} = 28$ V and $V_{Imax} = 32$ V. Plot all the three curves on the same figure window. Use `legend` or `text` commands to label the different curves clearly.

C. DC Voltage Transfer Function of the Buck–Boost Converter as a Function of the Duty Cycle

1. The equation for the dc voltage transfer function of the lossy buck–boost converter is given in Equation (15.7). For efficiency, use Equations (15.4) and (15.6).
2. Define the equation for the dc voltage transfer function on MATLAB® along with all the specifications and their values. Define a duty cycle range, that is, vary D from 0 to 1 in steps of 0.001.
3. Plot the dc voltage transfer function M_{VDC} as a function of the duty cycle for $R_L = R_{Lmin}$.

Post-lab Question

Consider Eq. 15.3. Replace the ratio I_O/V_O by $1/R_L$. The duty cycle D depends on the load resistance R_L. Plot the variation in the duty cycle D as a function of the normalized load resistance $R_L/(2f_sL)$. You may consider $R_L/(2f_sL)$ as a single term. Vary the term $R_L/(2f_sL)$ from 1 Ω to 1000 Ω. Use the logspace command to declare the variable. For example,

```
RL_2fs_L = logspace(0,3, 1000);
```

Using the `semilogx` command, plot the curve D vs. $R_L/(2f_sL)$.

16

Flyback DC–DC Converter in CCM—Steady-State Simulation

Objectives

The objectives of this lab are:

- To design a pulse-width modulated (PWM) flyback dc–dc converter operating in continuous-conduction mode (CCM) for the design specifications provided.
- To simulate and analyze the steady-state characteristics of the flyback converter in CCM.
- To determine the overall efficiency of the flyback converter.

Specifications

The specifications of the flyback converter are as provided in Table 16.1.

Pre-lab

For the specifications provided, find the values of all the components and parameters for the buck–boost dc–dc converter operating in CCM using the relevant design equations provided in Tables A.2 and A.3 in Appendix A.

Quick Design

Assume an overall efficiency of $\eta = 80\%$ and $D_{max} = 0.36$ resulting in turns ratio as $n = 11$. Choose:

$L_m = 2.5$ mH, ESR of the inductor $r_L = 0.35\ \Omega$, $R_{Lmin} = 0.5\ \Omega$, $R_{Lmax} = 5\ \Omega$, $D_{min} = 0.1555$, $D_{max} = 0.3638$ (or assumed value), $C = 4$ mF, ESR of the capacitor $r_C = 2.5$ mΩ.

Laboratory Manual for Pulse-Width Modulated DC–DC Power Converters, First Edition.
Marian K. Kazimierczuk and Agasthya Ayachit.
© 2016 John Wiley & Sons, Ltd. Published 2016 by John Wiley & Sons, Ltd.

Table 16.1   Parameters and their values

Parameter	Notation	Value
Minimum rms input voltage	$V_{Imin(rms)}$	85 V
Minimum dc input voltage	V_{Imin}	120.21 V
Maximum rms input voltage	$V_{Imax(rms)}$	264 V
Maximum dc input voltage	V_{Imax}	373.35 V
DC output voltage	V_O	5 V
Switching frequency	f_s	100 kHz
Maximum output current	I_{Omax}	10 A
Minimum output current	I_{Omin}	1 A
Output voltage ripple	V_r	$< 0.01 V_O$

MOSFET: International Rectifier IRF840 n-channel power MOSFET with $V_{DSS} = 500$ V, $I_{SM} = 8$ A, $r_{DS} = 0.85$ mΩ at $T = 25°C$, $C_o = 100$ pF, and $V_t = 4$ V.

Diode: ON Semiconductor MBR2540 with $V_{RRM} = 40$ V, $I_F = 25$ A, $R_F = 10$ mΩ, and $V_F = 0.3$ V.

Procedure

A. Analysis of the Flyback Converter in Steady State

1. Construct the circuit of the flyback converter shown in Figure 16.1 on the circuit simulator. You may use (a) a dc/dc ideal transformer and place an inductor L_m in parallel or (b) use the model of the two-winding transformer as shown at the end of this lab. Name all the nodes and components for convenience.
2. Initially, let the value of the input voltage be $V_{Imax} = 373.5$ V and the duty cycle be $D_{min} = 0.1555$. Set the value of the load resistance to $R_{Lmin} = 0.5$ Ω.
3. Enter the values of all the components. Place a pulse voltage source in order to provide the gate-to-source voltage at the MOSFET terminals. Set `time period = 10` μs, `duty cycle/width = 0.155`, and `amplitude = 12` V. Let the `rise time` and `fall time` be equal to zero (optional).

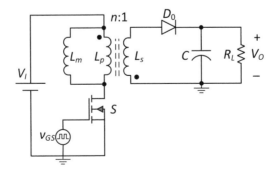

Figure 16.1   Circuit diagram of the PWM flyback dc–dc converter.

4. Set simulation type to `transient analysis`. Set `end time = 20` ms and a `time step = 0.1` μs. Run the simulation.
5. Plot the following parameters after successful completion of the simulation. You may display the waveforms on different figure windows for better clarity.
   - Gate-to-source voltage v_{GS}, drain-to-source voltage v_{DS}, and diode voltage v_D.
   - Output voltage v_O, output current i_O, and output power p_O.
   - Current through the magnetizing inductance i_{Lm}, diode current i_D, and MOSFET current i_S.
   Use the `zoom` option to display the steady-state region.
6. Observe the waveform of the current through the magnetizing inductance to ensure whether it is in CCM. If the current is not in CCM, then increase the value of the inductance L_m and repeat the simulation.
7. For the above-mentioned waveforms, measure:
   - The average and peak-to-peak values of the current through the magnetizing inductance L_m.
   - The maximum and average values of the voltage across the MOSFET S.
   - The minimum and average values of the voltage across the diode D_0.
   - The maximum and average values of the currents through the MOSFET and the diode.
   Ensure that the values obtained above match the desired specifications.
8. Repeat the steps above with duty cycle $D_{max} = 0.3638$ and input voltage $V_{Imin} = 120.21$ V.

B. Estimation of the Overall Efficiency of the Flyback Converter

1. Set up the converter to operate at nominal operating condition, that is, at $D = D_{nom}$ and $V_I = V_{Inom}$. Let $R_L = R_{Lmin}$ such that the converter delivers the maximum output power.
2. Set the simulation type to transient analysis and perform the simulation.
3. Plot the waveforms of the input power p_I and the output power p_O. Zoom in to the steady-state region.
4. Measure the average values of the input power and the output power. If the input power is negative, then consider only the magnitude of the average value of the input power.
5. Calculate the efficiency of the converter using $\eta = P_O/P_I$, where P_O is the average value of the output power and P_I is the average value of the input power, respectively.
6. This section may be repeated by plotting the power waveforms of all the components and then estimating their average values. All the power losses can be added to give the total power loss in the converter. Further, the efficiency can be estimated using $\eta = P_O/(P_{LS} + P_I)$, where P_O is the average value of the output power and P_{LS} is the sum of the average values of the power loss in individual components.

Post-lab Questions

1. Draw figures showing the transformation of a buck–boost converter into a flyback converter.
2. What are the basic differences between a buck–boost converter and a flyback converter?
3. What are the advantages of a flyback converter?
4. Mention a few practical applications of the flyback converter.
5. Is the flyback converter a step-down or a step-up converter?

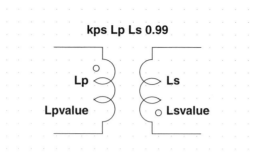

Figure 16.2   Circuit diagram of the PWM flyback dc–dc converter for Spice-based simulations.

6. What is the useful range of the dc voltage transfer function or duty cycle for the flyback converter?
7. Is the switch voltage stress low in the flyback converter? If not, suggest methods to suppress the maximum voltage stress across the switch.

Note

An example model of a two-winding transformer developed using PSpice is given as

```
1  * * * * * *
2  Lp              Node_1            Node_2            Lpvalue
3  Ls     Node_3   Node_4   Lsvalue
4  * * * * * *
```

where L_p represents the inductance of the primary winding and L_s represents the inductance of the secondary winding. The value of L_p given as Lpvalue is the chosen value of the magnetizing inductance needed to ensure CCM operation. The value of L_s given as Lsvalue is obtained as Lpvalue/n^2, where n is the turns ratio between the primary and secondary windings. A transformer on LTSpice can be constructed using the circuit shown in Figure 16.2.

A few simulation software such as PSpice and LTSpice restricts the user from plotting the waveform of the magnetizing inductance directly, since the magnetizing inductance is a nonphysical element. However, in software such as SABER and Cadence, a separate physical inductor can be placed in parallel with a dc/dc transformer and the current through the magnetizing inductance can be recorded.

17

Efficiency and DC Voltage Transfer Function of PWM Flyback DC–DC Converters in CCM

Objectives

The objectives of this lab are:

- To design a PWM flyback dc–dc converter in CCM using the design equations.
- To analyze the variations in efficiency of the lossy flyback converter at different loads and different input voltages.
- To analyze the variation in efficiency and lossy steady-state voltage transfer function.

Theory

The steady-state dc voltage transfer function of a converter is

$$M_{VDC} = \frac{V_O}{V_I}. \tag{17.1}$$

The overall converter efficiency is

$$\eta = \frac{N_\eta}{D_\eta}, \tag{17.2}$$

Laboratory Manual for Pulse-Width Modulated DC–DC Power Converters, First Edition.
Marian K. Kazimierczuk and Agasthya Ayachit.
© 2016 John Wiley & Sons, Ltd. Published 2016 by John Wiley & Sons, Ltd.

where

$$
N_\eta = 1 - M_{VDC}\left(\frac{2r_L + r_{DS} + r_{T1}}{nR_L}\right) - nM_{VDC}\left(\frac{R_F + r_{T2} + r_C}{R_L}\right)
$$

$$
+ \left\{ \left[1 - M_{VDC}\left(\frac{2r_L + r_{DS} + r_{T1}}{nR_L}\right) - nM_{VDC}\left(\frac{R_F + r_{T2} + r_C}{R_L}\right)\right]^2 \right.
$$

$$
\left. - \frac{4M_{VDC}^2(r_L + r_{DS} + r_{T1})}{R_L}\left[1 + \frac{r_L}{n^2 R_L} + \frac{R_F + r_{T2}}{R_L} + \frac{V_F}{V_O} + \frac{f_s C_o R_L(1 + nM_{VDC})^2}{M_{VDC}^2}\right] \right\}^{\frac{1}{2}},
$$

and

$$
D_\eta = 2\left[1 + \frac{r_L}{n^2 R_L} + \frac{R_F + r_{T2}}{R_L} + \frac{V_F}{V_O} + \frac{f_s C_o R_L(1 + nM_{VDC})^2}{M_{VDC}^2}\right],
$$

where r_L is the equivalent series resistance (ESR) of the magnetizing inductance, r_{DS} is the on-state resistance of the MOSFET, R_L is the load resistance, R_F is the forward resistance of the diode, r_C is the ESR of the output filter capacitor, V_F is the forward voltage drop of the diode, f_s is the switching frequency, and C_o is the output capacitance of the MOSFET, r_{T1} and r_{T2} are the ESR of the primary and secondary windings of the transformer, and n is the turns ratio between the primary and secondary windings.

The steady-state dc voltage transfer function of the lossy flyback converter is

$$
M_{VDC} = \frac{\eta D}{n(1 - D)}. \tag{17.3}
$$

Specifications

The specifications of the flyback converter are as provided in Table 17.1.

Table 17.1 Parameters and their values

Parameter	Notation	Value
Minimum rms input voltage	$V_{Imin(rms)}$	85 V
Minimum dc input voltage	V_{Imin}	120.21 V
Maximum rms input voltage	$V_{Imax(rms)}$	264 V
Maximum dc input voltage	V_{Imax}	373.35 V
DC output voltage	V_O	5 V
Switching frequency	f_s	100 kHz
Maximum output current	I_{Omax}	10 A
Minimum output current	I_{Omin}	1 A
Output voltage ripple	V_r	$< 0.01V_O$

Pre-lab

For the above specifications, find the values of all the components and specifications for the flyback dc–dc converter operating in CCM using the relevant design equations provided in Tables A.2 and A.3 in Appendix A.

Quick Design

Assume an overall efficiency of $\eta = 80\%$ and $D_{max} = 0.36$ resulting in turns ratio $n = 11$. Choose:

$L_m = 2.5$ mH, ESR of the inductor $r_L = 0.35\ \Omega$, $R_{Lmin} = 0.5\ \Omega$, $R_{Lmax} = 5\ \Omega$, $D_{min} = 0.1555$, $D_{max} = 0.3638$ (or assumed value), $C = 4$ mF, ESR of the capacitor $r_C = 2.5$ mΩ. Let $r_{T1} = 0.9\ \Omega$ and $r_{T2} = 0.02\ \Omega$.

MOSFET: International Rectifier IRF840 n-channel power MOSFET with $V_{DSS} = 500$ V, $I_{SM} = 8$ A, $r_{DS} = 0.85$ mΩ at $T = 25°C$, $C_o = 100$ pF, and $V_t = 4$ V.

Diode: ON Semiconductor MBR2540 with $V_{RRM} = 40$ V, $I_F = 25$ A, $R_F = 10$ mΩ, and $V_F = 0.3$ V.

Procedure

A. Efficiency of the Flyback Converter as a Function of the Input Voltage at Full, Nominal, and Light Load Conditions

1. The equation for the efficiency of the lossy flyback converter is given in Equation (17.2). Define the equation for the efficiency on MATLAB® along with all the specifications and their values. Let the term M_{VDC} in Equation (17.2) be equal to V_O/V_I. Initially, let $R_L = R_{Lmin} = V_O/I_{Omax}$.
2. Define an input voltage range, that is, vary V_I from 100 to 400 V in steps of 0.01 V.
3. Plot the efficiency as a function of the input voltage at $R_L = R_{Lmin}$.
4. Repeat the activity by replacing R_L with $R_{Lmax} = V_O/I_{Omin}$. Plot the two curves on the same figure window. Use `legend` or `text` commands to label the different curves clearly.

B. Efficiency of the Flyback Converter as a Function of the Output Current at Minimum, Nominal, and Maximum Input Voltages

1. Use the code developed in Section A. In this activity, let the value of M_{VDC} be fixed, that is, let $V_I = V_{Imin}$. Also, replace the term R_L with V_O/I_O.
2. Vary I_O from 1 to 10 A in steps of 0.001 A.
3. Plot the efficiency as a function of the output current at $V_I = V_{Imin}$.
4. Repeat the activity by replacing V_I with V_{Imax}. Plot the two curves on the same figure window. Use `legend` or `text` commands to label the different curves clearly.

C. DC Voltage Transfer Function of the Flyback Converter as a Function of the Duty Cycle

1. In the expression for efficiency in Equation (17.2), replace M_{VDC} with $D/[n(1-D)]$.

2. The equation for the dc voltage transfer function of the lossy flyback converter is given in Equation (17.3). Define the equation for the dc voltage transfer function on MATLAB® along with all the specifications and their values. Replace R_L with $R_{Lmin} = V_O/I_{Omax}$.
3. Define a duty cycle range, that is, vary D from 0 to 1 in steps of 0.001.
4. Plot the dc voltage transfer function M_{VDC} as a function of the duty cycle for $R_L = R_{Lmax}$.
5. Repeat the activity using $R_L = R_{Lmax}$. Plot the two curves on the same figure window. Use `legend` or `text` commands to label the different curves clearly.

Post-lab Question

Consider Eq. 17.2. Replace the term M_{VDC} by V_O/V_I. Plot the efficiency η as a function of the load resistance R_L for three different values of the input voltage. Vary R_L from 0.5 Ω to 5 Ω. Set the input voltage at $V_I = V_{Imin}$. Using the `plot` command, plot the curve η vs. R_L. Repeat the activity for $V_I = V_{Inom}$ and $V_I = V_{Imax}$. Present all the plots on the same figure window. Use `legend` or `text` commands to label the different curves clearly.

18

Multiple-Output Flyback DC–DC Converter in CCM

Objective

The objective of this lab is to simulate the pulse-width modulated (PWM) multiple-output flyback dc–dc converter operating in continuous-conduction mode (CCM) for two different voltage profiles.

Specifications

The specifications of the converter are as provided in Table 18.1.
The two different output voltage profiles are:

Case 1: $V_{O1} = V_{O2} = 12$ V.

Case 2: $V_{O1} = 15$ V, $V_{O2} = 5$ V.

Quick Design

Assume an overall efficiency of $\eta = 100\%$. Let $C_1 = C_2 = 220$ μF.

MOSFET: International Rectifier IRF840 n-channel power MOSFET with $V_{DSS} = 500$ V, $I_{SM} = 8$ A, $r_{DS} = 0.85$ mΩ at $T = 25°$C, $C_o = 100$ pF, and $V_t = 4$ V.

Diode: ON Semiconductor MBR2540 with $V_{RRM} = 40$ V, $I_F = 25$ A, $R_F = 10$ mΩ, and $V_F = 0.3$ V.

Procedure

A. Simulation of the Multiple-Output Flyback Converter

1. Construct the circuit of the flyback converter shown in Figure 18.1 on the circuit simulator. Details about constructing the transformer has been discussed in Lab 16.

Laboratory Manual for Pulse-Width Modulated DC–DC Power Converters, First Edition.
Marian K. Kazimierczuk and Agasthya Ayachit.
© 2016 John Wiley & Sons, Ltd. Published 2016 by John Wiley & Sons, Ltd.

Table 18.1   Parameters and their values

Parameter	Notation	Value
Minimum dc input voltage	V_{Imin}	20 V
Maximum dc input voltage	V_{Imax}	50 V
Switching frequency	f_s	100 kHz
Maximum output power	P_{Omax}	75 W
Minimum output current	P_{Omin}	10 W
Output voltage ripple	V_r	$< 0.01V_O$

2. Initially, consider the component values for Case 1 provided in Table 18.2. Also, consider the minimum value of the duty cycle D_{min} and correspondingly the maximum value of the input voltage V_{Imax}. Enter the values of all the other components.
3. Connect a pulse voltage source to the gate and source terminals of the MOSFET. Set `time period = 10 μs`, `duty cycle/width = 0.2466`, and `amplitude = 12 V`. Let the `rise time` and `fall time` be equal to zero (optional).
4. Set simulation type to `transient analysis`. Set `end time = 20 ms` and `time step = 0.1 μs`. Run the simulation.
5. Plot the following parameters after successful completion of the simulation. You may display the waveforms on different figure windows for better clarity.
   - Gate-to-source voltage v_{GS} and drain-to-source voltage v_{DS}.
   - Current through the magnetizing inductance i_{Lm} and voltage across the magnetizing inductance v_{Lm}.
   - The currents through the diodes D_1 and D_2 and the current through the MOSFET.
   - The voltages across the diodes D_1 and D_2.
   - Output voltage V_O, output current I_O, and output power P_O.
   Use the `zoom` option to focus only on the steady-state region.
6. Observe the waveform of the current through the magnetizing inductance to ensure whether it is in CCM. If the current is not in CCM, then increase the value of the inductance L_m and repeat the simulation.
7. Repeat the above steps for the set of specifications presented for Case 2 using the component values shown in Table 18.2.

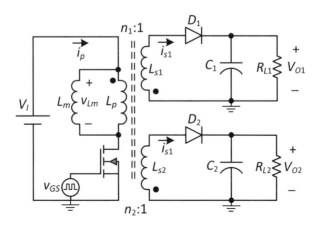

Figure 18.1   Circuit diagram of the multiple-output flyback dc–dc converter.

Table 18.2 Component values for the different cases

Case	Magnetizing Inductance	n_1	n_2	R_{L1}	R_{L2}
Case 1	39 μH	1.22	1.22	14.40 Ω	14.40 Ω
Case 2	39 μH	0.98	2.94	22.5 Ω	2.5 Ω

Post-lab Questions

1. What are the areas of application of multiple-output flyback converters?
2. What are the advantages of multiple-output flyback converters?
3. What are the disadvantages of multiple-output flyback converters?
4. How does the value of the magnetizing inductance vary with an increase in the number of outputs?

19

Flyback DC–DC Converter in DCM—Steady-State Simulation

Objectives

The objectives of this lab are:

- To design a pulse-width modulated (PWM) flyback dc–dc converter operating in discontinuous-conduction mode (DCM) for the design specifications provided.
- To simulate and analyze the steady-state characteristics of the flyback converter in DCM.
- To determine the overall efficiency of the flyback converter.

Specifications

The specifications of the flyback converter are as provided in Table 19.1.

Pre-lab

For the specifications provided, find the values of all the components and specifications for the flyback dc–dc converter operating in DCM using the relevant design equations provided in Tables B.2 and B.3 in Appendix B.

Quick Design

Assume an overall efficiency of $\eta = 85\%$ and $D_{Bmax} = 0.4$ resulting in turns ratio $n = 5$. Choose:

$L_m = 300$ μH, ESR of the inductor $r_L = 0.5$ Ω, $R_{Lmin} = 7.5$ Ω, $D_{min} = 0.123$, $D_{max} = 0.383$ (or assumed value), $C = 120$ μF, ESR of the capacitor $r_C = 10$ mΩ.

Laboratory Manual for Pulse-Width Modulated DC–DC Power Converters, First Edition.
Marian K. Kazimierczuk and Agasthya Ayachit.
© 2016 John Wiley & Sons, Ltd. Published 2016 by John Wiley & Sons, Ltd.

Table 19.1   Parameters and their values

Parameter	Notation	Value
Minimum rms input voltage	$V_{Imin(rms)}$	85 V
Minimum dc input voltage	V_{Imin}	120.21 V
Maximum rms input voltage	$V_{Imax(rms)}$	264 V
Maximum dc input voltage	V_{Imax}	373.35 V
DC output voltage	V_O	15 V
Switching frequency	f_s	100 kHz
Maximum output current	I_{Omax}	2 A
Minimum output current	I_{Omin}	0 A
Output voltage ripple	V_r	$< 0.01V_O$

MOSFET: International Rectifier IRF840 n-channel power MOSFET with $V_{DSS} = 500$ V, $I_{SM} = 8$ A, $r_{DS} = 0.85$ mΩ at $T = 25°C$, $C_o = 100$ pF, and $V_t = 4$ V.

Diode: ON Semiconductor MBR10100 Schottky diode with $V_{RRM} = 100$ V, $I_F = 10$ A, $R_F = 30$ mΩ, and $V_F = 0.35$ V.

Procedure

A. Simulation of the Flyback Converter in DCM and its Analysis in Steady State

1. Construct the circuit of the flyback converter shown in Figure 19.1 on the circuit simulator. Name all the nodes and components for convenience.
2. Initially, let the value of the input voltage be $V_{Imax} = 373.5$ V and the duty cycle be $D_{min} = 0.1555$. Set the value of the load resistance to $R_{Lmin} = 7.5$ Ω.
3. Enter the values of all the components. Place a pulse voltage source in order to provide the gate-to-source voltage at the MOSFET terminals. Set `time period = 10` μs, `duty cycle/width = 0.123`, and `amplitude = 12` V. Let the `rise time` and `fall time` be equal to zero (optional). Let the input voltage V_I be equal to $V_{Imax} = 373.5$ V.
4. Set simulation type to `transient analysis`. Set `end time = 10` ms and a `time step = 0.1` μs. Run the simulation.

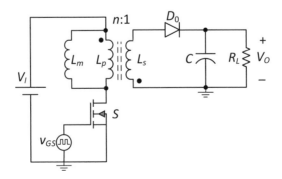

Figure 19.1   Circuit diagram of the PWM flyback dc–dc converter.

5. Plot the following parameters after successful completion of the simulation. You may display the waveforms on different figure windows for better clarity.
   - Gate-to-source voltage v_{GS}, drain-to-source voltage v_{DS}, and diode voltage v_D.
   - Output voltage v_O, output current i_O, and output power p_O.
   - Current through the magnetizing inductance i_{Lm}, diode current i_D, and MOSFET current i_S.

   Use the zoom option to display only the steady-state region.
6. Observe the inductor current waveform to ensure whether the current is in DCM. If the current is not in DCM, then decrease the value of the inductor by a small value and repeat the simulation.
7. For the above-mentioned waveforms, measure:
   - The average and peak-to-peak values of the current through the magnetizing inductance L_m.
   - The maximum, intermediate, and average values of the voltage across the MOSFET S.
   - The minimum, intermediate, and average values of the voltage across the diode D_0.
   - The maximum and average values of the currents through the MOSFET and the diode.

   Ensure that the values obtained above match the desired specifications.
8. Repeat the steps above with duty cycle $D_{max} = 0.383$ and input voltage $V_{Imin} = 120.21$ V.

B. Estimation of the Overall Efficiency of the Flyback Converter

1. Set up the converter to operate at nominal operating condition, that is, at $D = D_{nom}$ and $V_I = V_{Inom}$. Let $R_L = R_{Lmin}$ such that the converter delivers the maximum output power.
2. Set the simulation type to transient analysis and perform the simulation.
3. Plot the waveforms of the input power p_I and the output power p_O. Zoom in to the steady-state region.
4. Measure the average values of the input power and the output power. If the input power is negative, then consider only the magnitude of the average value of the input power.
5. Calculate the efficiency of the converter using $\eta = P_O/P_I$, where P_O is the average value of the output power and P_I is the average value of the input power, respectively.
6. This section may be repeated by plotting the power waveforms of all the components and then estimating their average values. All the power losses can be added to give the total power loss in the converter. Further, the efficiency can be estimated using $\eta = P_O/(P_{LS} + P_I)$, where P_O is the average value of the output power and P_{LS} is the sum of the average values of the power loss in individual components.

Post-lab Questions

1. Mention a few applications where the flyback converters are operated in the discontinuous conduction mode.
2. Draw the waveform of the current flowing through the magnetizing inductance, when the converter operates in the discontinuous-conduction mode and at maximum input voltage.

20

Efficiency and DC Voltage Transfer Function of PWM Flyback DC–DC Converter in DCM

Objectives

The objectives of this lab are:

- To design a PWM flyback dc–dc converter in DCM using the design equations.
- To analyze the variations in efficiency of the lossy flyback converter in DCM at different load resistances and different input voltages.
- To observe the dependence of the lossy dc voltage transfer function on duty cycle.

Theory

The steady-state dc voltage transfer function of a converter is

$$M_{VDC} = \frac{V_O}{V_I}. \tag{20.1}$$

In terms of the circuit parameters, the transfer function under ideal conditions can be expressed as

$$M_{VDC} = nD\sqrt{\frac{V_O}{2f_s L_m I_O}}, \tag{20.2}$$

resulting in

$$D = 1 - \sqrt{\frac{2f_s L_m I_O}{n^2 V_O}}. \tag{20.3}$$

Laboratory Manual for Pulse-Width Modulated DC–DC Power Converters, First Edition.
Marian K. Kazimierczuk and Agasthya Ayachit.
© 2016 John Wiley & Sons, Ltd. Published 2016 by John Wiley & Sons, Ltd.

The overall converter efficiency is

$$\eta = \cfrac{1}{1 + \cfrac{P_{LS}}{P_O}}, \tag{20.4}$$

where P_{LS} is the total power loss in the converter in terms of M_{VDC} is

$$P_{LS} = \left[\frac{2M_{VDC}(r_{DS} + r_{T1})}{3} \sqrt{\frac{2}{f_s L_m R_L}} + \frac{2n(R_F + r_{T2})}{3} \sqrt{\frac{2}{f_s L_m R_L}} \right.$$
$$\left. + \frac{2r_L M_{VDC}}{3} \sqrt{\frac{2}{f_s L_m R_L}} \left(1 + \frac{1}{nM_{VDC}}\right) + \frac{V_F}{V_O} + n^2 f_s C_o R_L \left(\frac{1}{nM_{VDC}} + 1\right)^2 \right] P_O, \tag{20.5}$$

where r_L is the equivalent series resistance (ESR) of the magnetizing inductance, r_{DS} is the on-state resistance of the MOSFET, R_L is the load resistance, R_F is the forward resistance of the diode, r_C is the ESR of the output filter capacitor, V_F is the forward voltage drop of the diode, f_s is the switching frequency, and C_o is the output capacitance of the MOSFET, r_{T1} and r_{T2} are the ESR of the primary and secondary windings of the transformer, and n is the turns ratio between the primary and secondary windings. In terms of the duty cycle, the total power loss P_{LS} can be expressed as

$$P_{LS} = \left[\frac{2D(r_{DS} + r_{T1})}{3f_s L_m} + \frac{2n(R_F + r_{T2})}{3} \sqrt{\frac{2}{f_s L_m R_L}} + \frac{2Dr_L}{3f_s L_m} \left(1 + \sqrt{\frac{2f_s L_m}{n^2 D^2 R_L}}\right) \right.$$
$$\left. + \frac{V_F}{V_O} + n^2 f_s C_o R_L \left(\sqrt{\frac{2f_s L_m}{n^2 D^2 R_L}} + 1\right)^2 \right] P_O. \tag{20.6}$$

The steady-state dc voltage transfer function of the lossy flyback converter in DCM is

$$M_{VDC} = nD\sqrt{\frac{\eta R_L}{2f_s L_m}} = nD\sqrt{\frac{\eta V_O}{2f_s L_m I_O}}. \tag{20.7}$$

Specifications

The specifications of the flyback converter are as provided in Table 20.1

Pre-lab

For the specifications provided, find the values of all the components and parameters for the flyback dc–dc converter operating in DCM using the relevant design equations provided in Tables B.2 and B.3 in Appendix B.

Table 20.1 Parameters and their values

Parameter	Notation	Value
Minimum rms input voltage	$V_{Imin(rms)}$	85 V
Minimum dc input voltage	V_{Imin}	120.21 V
Maximum rms input voltage	$V_{Imax(rms)}$	264 V
Maximum dc input voltage	V_{Imax}	373.35 V
DC output voltage	V_O	15 V
Switching frequency	f_s	100 kHz
Maximum output current	I_{Omax}	2 A
Minimum output current	I_{Omin}	0 A
Output voltage ripple	V_r	$< 0.01V_O$

Quick Design

Assume an overall efficiency of $\eta = 85\%$ and $D_{Bmax} = 0.4$ resulting in turns ratio $n = 5$. Choose:

$L_m = 300\ \mu H$, ESR of the inductor $r_L = 0.5\ \Omega$, $R_{Lmin} = 7.5\ \Omega$, $D_{min} = 0.123$, $D_{max} = 0.383$ (or assumed value), $C = 120\ \mu F$, ESR of the capacitor $r_C = 10\ m\Omega$.

MOSFET: International Rectifier IRF840 n-channel power MOSFET with $V_{DSS} = 500$ V, $I_{SM} = 8$ A, $r_{DS} = 0.85\ m\Omega$ at $T = 25°C$, $C_o = 100$ pF, and $V_t = 4$ V.

Diode: ON Semiconductor MBR10100 Schottky diode with $V_{RRM} = 100$ V, $I_F = 10$ A, $R_F = 30\ m\Omega$, and $V_F = 0.35$ V.

Procedure

A. Efficiency of the Flyback Converter as a Function of the Input Voltage at Different Load Conditions

1. The equations for the efficiency and total power loss of the lossy flyback converter are given in Equations (20.4) and (20.5). In Equation (20.5), replace the term M_{VDC} with V_O/V_I. Initially, let $R_L = 75\ \Omega$.
2. Define the equation for the efficiency using Equations (20.4) and (20.5) on MATLAB® along with all the specifications and their values. Define an input voltage range, that is, vary V_I from 100 to 400 V in steps of 1 V.
3. Plot the efficiency as a function of the input voltage for $R_L = 75\ \Omega$.
4. Repeat the activity for $R_L = 7.5\ \Omega$ and $R_L = 15\ \Omega$. Plot all the three curves on the same figure window. Use `legend` or `text` commands to label the different curves clearly.

B. Efficiency of the Flyback Converter as a Function of the Output Current at Minimum, Nominal, and Maximum Input Voltages

1. Use the MATLAB® code developed in Section A. Let M_{VDC} be fixed and be equal to $V_I = V_{Imin} = 120.21$ V. Replace the term R_L with V_O/I_O.

2. Define the equation for the efficiency on MATLAB® along with all the specifications and their values. Define a range for the output current, that is, vary I_O from 0 to 2 A in steps of 0.001 A.
3. Plot the efficiency as a function of the output current for $V_I = V_{Imin}$.
4. Repeat the activity for $V_{Imax} = 373.35$ V. Plot the two curves on the same figure window. Use legend or text commands to label the different curves clearly.

C. DC Voltage Transfer Function of the Flyback Converter as a Function of the Duty Cycle

1. The equation for the dc voltage transfer function of the lossy flyback converter is given in Equation (20.7). Use the expression given in Equation (20.6) for the total power loss in terms of duty cycle.
2. Define the equation for the dc voltage transfer function on MATLAB® along with all the specifications and their values. Define a duty cycle range, that is, vary D from 0 to 1 in steps of 0.001.
3. Plot the dc voltage transfer function M_{VDC} as a function of the duty cycle for $R_L = R_{Lmin}$.

Post-lab Question

Explain the significance of all the plots obtained in Sections A, B, and C.

21

Forward DC–DC Converter in CCM—Steady-State Simulation

Objectives

The objectives of this lab are:

- To design a pulse-width modulated (PWM) forward dc–dc converter operating in continuous-conduction mode (CCM) for the design specifications provided.
- To simulate and analyze the steady-state characteristics of the forward converter in CCM.
- To determine the overall efficiency of the forward converter.

Specifications

The specifications of the forward converter are as provided in Table 21.1.

Pre-lab

For the above specifications, find the values of all the components and parameters of the forward dc–dc converter operating in CCM using the relevant design equations provided in Tables A.2 and A.3 in Appendix A.

Quick Design

Assume an overall efficiency of $\eta = 80\%$ and $D_{max} = 0.4$. The turns ratio between primary and secondary windings $n_1 = 8$ and the turns ratio between the primary and tertiary (reset) windings $n_3 = 8$. Choose:

$L_m = 2\,\text{mH}, L = 20\,\mu\text{H}$, ESR of the inductor L is $r_L = 0.015\,\Omega, R_{Lmin} = 0.25\,\Omega, R_{Lmax} = 2.5\,\Omega,$ $D_{min} = 0.2674, \ D_{nom} = 0.3205, \ D_{max} = 0.3937$ (or assumed value), $C = 200\,\mu\text{F}$, ESR of

Laboratory Manual for Pulse-Width Modulated DC–DC Power Converters, First Edition.
Marian K. Kazimierczuk and Agasthya Ayachit.

Table 21.1 Parameters and component values

Parameters	Notation	Value
Minimum dc input voltage	V_{Imin}	127 V
Nominal dc input voltage	V_{Inom}	156 V
Maximum dc input voltage	V_{Imax}	187 V
DC output voltage	V_O	5 V
Switching frequency	f_s	100 kHz
Maximum output current	I_{Omax}	20 A
Minimum output current	I_{Omin}	2 A
Output voltage ripple	V_r	$< 0.01 V_O$

the capacitor $r_C = 25$ mΩ, ESR of the primary and secondary windings are $r_{T1} = 50$ mΩ, $r_{T2} = 10$ mΩ, respectively.

MOSFET: International Rectifier IRF740 n-channel power MOSFET with $V_{DSS} = 400$ V, $I_{SM} = 10$ A, $r_{DS} = 0.55$ Ω at $T = 25°C$, $C_o = 100$ pF, and $V_t = 4$ V.

Diode: Two ON Semiconductor MBR2540 diodes for D_1 and D_2 with $V_{RRM} = 40$ V, $I_F = 25$ A, $R_F = 10$ mΩ, and $V_F = 0.3$ V and a MR826 fast recovery diode for D_3 with $V_{RRM} = 600$ V, $I_{F(AV)} = 35$ A. $R_F = 10$ mΩ, and $V_F = 1.2$ V.

Procedure

A. Simulation of the Forward Converter in Steady State

1. Construct the circuit of the forward converter shown in Figure 21.1 on the circuit simulator. The circuit of the transformer, which can be used for this converter has been discussed in Lab 16. Choose an appropriate number of turns. You may select $N_1 = 16$ and $N_2 = N_3 = 2$.
2. Initially, let the value of the input voltage be $V_{Imax} = 187$ V and the duty cycle be $D_{min} = 0.2674$. Set the value of the load resistance to $R_{Lmin} = 0.25$ Ω.
3. Enter the values of all the components. Place a pulse voltage source in order to provide the gate-to-source voltage at the MOSFET terminals. Set `time period = 10` μs, `duty cycle/width = 0.2674`, and `amplitude = 12` V. Let the `rise time` and `fall time` be equal to zero (optional).

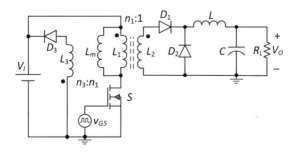

Figure 21.1 Circuit diagram of the PWM forward dc–dc converter.

4. Set simulation type to `transient analysis`. Set `end time = 20 ms` and a `time step = 0.1 μs`. Run the simulation.
5. Plot the following parameters after successful completion of the simulation. You may display the waveforms on different figure windows for better clarity.
   - Gate-to-source voltage v_{GS}, drain-to-source voltage v_{DS}, and diode voltages v_{D1}, v_{D2}, and v_{D3}.
   - Output voltage V_O, output current I_O, and output power P_O.
   - Current through the magnetizing inductance i_{Lm}, current through the output filter inductance i_L, diode currents i_{D1}, i_{D2}, i_{D3}, and MOSFET current i_S.
   Use the `zoom` option to focus only on the steady-state region.
6. Observe the waveform of the current through the output filter inductance to ensure whether it is in CCM. If the current is not in CCM, then increase the value of the inductance L and repeat the simulation.
7. For the above-mentioned waveforms, measure:
   - The average and peak-to-peak values of the current through the magnetizing inductance and the output filter inductance.
   - The maximum and average values of the voltage across the MOSFET S.
   - The minimum and average values of the voltage across the diodes D_1, D_2, and D_3.
   - The maximum and average values of the currents through the MOSFET and the diodes.
   Ensure that the values obtained above match the desired specifications.
8. Repeat the above steps by changing the duty cycle to $D_{max} = 0.3937$ and the input voltage to $V_{Imin} = 127$ V. Repeat for $D_{nom} = 0.3205$ and $V_{Inom} = 156$ V.

B. Estimation of the Overall Efficiency of the Forward Converter

1. Set up the converter to operate at nominal operating condition, that is, at $D = D_{nom}$ and $V_I = V_{Inom}$. Let $R_L = R_{Lmin}$ such that the converter delivers the maximum output power.
2. Set the simulation type to transient analysis and perform the simulation.
3. Plot the waveforms of the input power p_I and the output power p_O. Zoom in to the steady-state region.
4. Measure the average values of the input power and the output power. If the input power is negative, then consider only the magnitude of the average value of the input power.
5. Calculate the efficiency of the converter using $\eta = P_O/P_I$, where P_O is the average value of the output power and P_I is the average value of the input power, respectively.
6. This section may be repeated by plotting the power waveforms of all the components and then estimating their average values. All the power losses can be added to give the total power loss in the converter. Further, the efficiency can be estimated using $\eta = P_O/(P_{LS} + P_I)$, where P_O is the average value of the output power and P_{LS} is the sum of the average values of the power loss in individual components.

Post-lab Questions

1. Mention a few differences between the forward and flyback converters.
2. Is the magnetizing inductance in the forward converter required to store energy?

3. Explain the operation of the forward converter in CCM.

4. What is the purpose of the reset (or tertiary) winding in the forward converter?

Note

An example model of a two-winding transformer developed using PSpice can be obtained from Lab 16. The circuit can be modified into a three-winding transformer by adding a tertiary winding and including the component name in the code for coefficient coupling.

22

Efficiency and DC Voltage Transfer Function of PWM Forward DC–DC Converter in CCM

Objectives

The objectives of this lab are:

- To design a PWM forward dc–dc converter in CCM using the design equations.
- To analyze the variations in efficiency of the lossy forward converter in CCM at different load resistances and different input voltages.
- To observe the dependence of the lossy dc voltage transfer function on duty cycle.

Theory

The steady-state dc voltage transfer function of a converter is

$$M_{VDC} = \frac{V_O}{V_I}. \tag{22.1}$$

The overall converter efficiency is

$$\eta = \frac{N_\eta}{D_\eta}, \tag{22.2}$$

Laboratory Manual for Pulse-Width Modulated DC–DC Power Converters, First Edition.
Marian K. Kazimierczuk and Agasthya Ayachit.
© 2016 John Wiley & Sons, Ltd. Published 2016 by John Wiley & Sons, Ltd.

where

$$N_\eta = 1 - n_1 M_{VDC} \left(\frac{r_{DS} + r_{T1}}{n_1^2 R_L} + \frac{r_{T2}}{R_L} - \frac{r_C R_L}{6f_s^2 L^2} \right)$$

$$+ \left\{ \left[n_1 M_{VDC} \left(\frac{r_{DS} + r_{T1}}{n_1^2 R_L} + \frac{r_{T2}}{R_L} - \frac{r_C R_L}{6f_s^2 L^2} \right) \right]^2 \right.$$

$$\left. - \left[\frac{R_L r_C n_1^2 M_{VDC}^2}{3f_s^2 L^2} \left(1 + \frac{V_F}{V_O} + \frac{r_L + R_F}{R_L} + \frac{f_s C_o R_L}{M_{VDC}^2} + \frac{r_C R_L}{12f_s^2 L^2} \right) \right] \right\}^{\frac{1}{2}}, \quad (22.3)$$

and

$$D_\eta = 2 \left(1 + \frac{V_F}{V_O} + \frac{R_F + r_L}{R_L} + \frac{f_s C_o R_L}{M_{VDC}^2} + \frac{r_C R_L}{12f_s^2 L^2} \right), \quad (22.4)$$

where r_L is the equivalent series resistance (ESR) of the output filter inductor, r_{DS} is the on-state resistance of the MOSFET, R_L is the load resistance, R_F is the forward resistance of the diode, r_C is the ESR of the output filter capacitor, V_F is the forward voltage drop of the diode, f_s is the switching frequency, and C_o is the output capacitance of the MOSFET, r_{T1} and r_{T2} are the ESR of the primary and secondary windings of the transformer, and n_1 is the turns ratio between the primary and secondary windings.

The steady-state dc voltage transfer function of the lossy forward converter is

$$M_{VDC} = \frac{\eta D}{n_1}. \quad (22.5)$$

Specifications

The specifications of the buck converter are as provided in Table 22.1

Pre-lab

For the above specifications, find the values of all the components and parameters for the forward dc–dc converter operating in CCM using the relevant design equations provided in Tables A.2 and A.3 in Appendix A.

Quick Design

Assume an overall efficiency of $\eta = 80\%$ and $D_{max} = 0.4$. The turns ratio between primary and secondary windings $n_1 = 8$ and the turns ratio between the primary and tertiary (reset) windings $n_3 = 8$. Choose:

$L_m = 2\,\text{mH}, L = 20\,\mu\text{H}$, ESR of the inductor L is $r_L = 0.015\,\Omega, R_{Lmin} = 0.25\,\Omega, R_{Lmax} = 2.5\,\Omega$, $D_{min} = 0.2674, D_{nom} = 0.3205, D_{max} = 0.3937$ (or assumed value), $C = 200\,\mu\text{F}$, ESR of

Table 22.1 Parameters and their values

Parameter	Notation	Value
Minimum rms input voltage	$V_{Imin(rms)}$	90 V
Minimum dc input voltage	V_{Imin}	127 V
Nominal rms input voltage	$V_{Inom(rms)}$	110 V
Nominal dc input voltage	V_{Inom}	156 V
Maximum rms input voltage	$V_{Imax(rms)}$	132 V
Maximum dc input voltage	V_{Imax}	187 V
DC output voltage	V_O	5 V
Switching frequency	f_s	100 kHz
Maximum output current	I_{Omax}	20 A
Minimum output current	I_{Omin}	2 A
Output voltage ripple	V_r	$< 0.01V_O$

the capacitor $r_C = 25$ mΩ, ESR of the primary and secondary windings are $r_{T1} = 50$ mΩ, $r_{T2} = 10$ mΩ, respectively.

MOSFET: International Rectifier IRF740 n-channel power MOSFET with $V_{DSS} = 400$ V, $I_{SM} = 10$ A, $r_{DS} = 0.55$ mΩ at $T = 25°C$, $C_o = 100$ pF, and $V_t = 4$ V.

Diode: Two ON Semiconductor MBR2540 diodes for D_1 and D_2 with $V_{RRM} = 40$ V, $I_F = 25$ A, $R_F = 10$ mΩ, and $V_F = 0.3$ V and a MR826 fast recovery diode for D_3 with $V_{RRM} = 600$ V, $I_{F(AV)} = 35$ A. $R_F = 10$ mΩ, and $V_F = 1.2$ V.

Procedure

A. Efficiency of the Forward Converter as a Function of the Input Voltage at Full and Light Load Conditions

1. The equation for the efficiency of the lossy forward converter is given in Equation (22.2). Replace the term M_{VDC} in Equations (22.3) and (22.4) with V_O/V_I. Initially, let $R_L = R_{Lmin} = V_O/I_{Omax}$.
2. Define the equation for the efficiency on MATLAB® along with all the specifications and their values. Define an input voltage range, that is, vary V_I from 90 to 220 V in steps of 0.01 V.
3. Plot the efficiency as a function of the input voltage $R_L = R_{Lmin}$.
4. Repeat the activity by replacing R_L with $R_{Lmax} = V_O/I_{Omin}$. Plot the two curves on the same figure window. Use legend or text commands to label the different curves clearly.

B. Efficiency of the Forward Converter as a Function of the Output Current at Minimum, Nominal, and Maximum Input Voltages

1. Use the code developed in Section A. In this activity, let the value of M_{VDC} be fixed, that is, let $V_I = V_{Imin}$. Also, replace the term R_L with V_O/I_O.
2. Vary I_O from 1 to 10 A in steps of 0.001 A.

3. Plot the efficiency as a function of the output current at $V_I = V_{Imin}$.
4. Repeat the activity by replacing V_I with V_{Inom} and V_{Imax}. Plot all the three curves on the same figure window. Use `legend` or `text` commands to label the different curves clearly.

C. DC Voltage Transfer Function of the Forward Converter as a Function of the Duty Cycle

1. In the expression for the efficiency in Equations (22.3) and (22.4), replace M_{VDC} with D/n_1.
2. The equation for the dc voltage transfer function of the lossy forward converter is given in Equation (22.5). Define the equation for the dc voltage transfer function on MATLAB® along with all the specifications and their values. Replace R_L with $R_{Lmin} = V_O/I_{Omax}$.
3. Define a duty cycle range, that is, vary D from 0 to 1 in steps of 0.001.
4. Plot the dc voltage transfer function M_{VDC} as a function of the duty cycle for $R_L = R_{Lmin}$.
5. Repeat the activity using $R_L = R_{Lmax}$. Plot the two curves on the same figure window. Use `legend` or `text` commands to label the different curves clearly.

Post-lab Question

Explain the significance of all the plots obtained in Sections A, B, and C.

23

Forward DC–DC Converter in DCM—Steady-State Simulation

Objectives

The objectives of this lab are:

- To design a pulse-width modulated forward dc–dc converter operating in discontinuous-conduction mode (DCM) for the design specifications provided.
- To simulate the converter and analyze its characteristics in steady state.
- To estimate the overall efficiency of the forward converter in DCM.

Specifications

The specifications of the forward converter are given in Table 23.1.

Pre-lab

For the specifications provided, find the values of all the components and parameters for the forward dc–dc converter operating in DCM using the relevant design equations provided in Appendix B.

Quick Design

Assume an overall efficiency of $\eta = 80\%$ and $D_{max} = 0.4$. The turns ratio between primary and secondary windings $n_1 = 8$ and the turns ratio between the primary and tertiary (reset) windings $n_3 = 8$. Choose:

Laboratory Manual for Pulse-Width Modulated DC–DC Power Converters, First Edition.
Marian K. Kazimierczuk and Agasthya Ayachit.
© 2016 John Wiley & Sons, Ltd. Published 2016 by John Wiley & Sons, Ltd.

Table 23.1 Parameters and component values

Parameters	Notation	Value
Minimum rms input voltage	$V_{Imin(rms)}$	85 V
Minimum dc input voltage	V_{Imin}	120.21 V
Maximum rms input voltage	$V_{Imax(rms)}$	264 V
Maximum dc input voltage	V_{Imax}	373.35 V
DC output voltage	V_O	5 V
Switching frequency	f_s	100 kHz
Maximum output current	I_{Omax}	20 A
Minimum output current	I_{Omin}	0 A
Output voltage ripple	V_r	$< 0.01 V_O$

$L_m = 0.5$ mH, $L = 0.56$ μH, ESR of the inductor L is $r_L = 0.015$ Ω, $R_{Lmin} = 0.25$ Ω, $D_{min} = 0.0849$, $D_{max} = 0.3055$ (or assumed value), $C = 1$ mF, ESR of the capacitor $r_C = 7$ mΩ, ESR of the primary and secondary windings are $r_{T1} = 50$ mΩ, $r_{T2} = 10$ mΩ, respectively.

MOSFET: International Rectifier IRF740 n-channel power MOSFET with $V_{DSS} = 400$ V, $I_{SM} = 10$ A, $r_{DS} = 0.55$ Ω at $T = 25°C$, $C_o = 100$ pF, and $V_t = 4$ V.

Diode: Two ON Semiconductor MBR2540 diodes for D_1 and D_2 with $V_{RRM} = 40$ V, $I_F = 25$ A, $R_F = 10$ mΩ, and $V_F = 0.3$ V and a fast recovery MR826 fast recovery diode for D_3 with $V_{RRM} = 600$ V, $I_{F(AV)} = 35$ A. $R_F = 10$ mΩ, and $V_F = 1.2$ V.

Procedure

A. Simulation of the Forward Converter in Steady State in DCM

1. Construct the circuit of the forward converter shown in Figure 23.1 on the circuit simulator. Name all the nodes and components for convenience.
2. Initially, let the value of the input voltage be $V_{Imax} = 373.35$ V and the duty cycle be $D_{min} = 0.0849$. Set the value of the load resistance to $R_{Lmin} = 0.25$ Ω.
3. Enter the values of all the components. Place a pulse voltage source in order to provide the gate-to-source voltage at the MOSFET terminals. Set `time period = 10` μs, `duty cycle/width = 0.0849`, and `amplitude = 12` V. Let the `rise time` and `fall time` be equal to zero (optional).

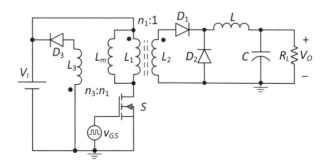

Figure 23.1 Circuit diagram of the PWM forward dc–dc converter.

4. Set simulation type to `transient analysis`. Set `end time = 10 ms` and `time step = 0.1 μs`. Run the simulation.
5. Plot the following parameters after successful completion of the simulation. You may display the waveforms on different figure windows for better clarity.
   - Gate-to-source voltage v_{GS}, drain-to-source voltage v_{DS}, and diode voltage v_D.
   - Output voltage V_O, output current I_O, and output power P_O.
   - Inductor current i_L, diode current i_D, and MOSFET current i_S.
     Use the `zoom` option to display only the steady-state region.
6. Observe the inductor current waveform to ensure whether the current is in DCM. If the current in not in DCM, then decrease the value of the inductor L and repeat the simulation.
7. For the above-mentioned waveforms, measure:
   - The average and peak-to-peak values of the current through the inductor L.
   - The maximum, intermediate, and average values of the voltage across the MOSFET S.
   - The minimum, intermediate, and average values of the voltage across the diode D_0.
   - The maximum and average values of the currents through the MOSFET and the diode.
     Ensure that the values obtained above match the desired specifications.
8. Repeat the steps above with duty cycle $D_{max} = 0.3055$ and input voltage $V_{Imin} = 120.21$ V.

B. Estimation of the Overall Efficiency

1. Set up the converter to operate at nominal operating condition, that is, at $D = D_{nom}$ and $V_I = V_{Inom}$. Let $R_L = R_{Lmin}$ such that the converter delivers the maximum output power.
2. Set the simulation type to transient analysis and perform the simulation.
3. Plot the waveforms of the input power p_I and the output power p_O. Zoom in to the steady-state region.
4. Measure the average values of the input power and the output power. If the input power is negative, then consider only the magnitude of the average value of the input power.
5. Calculate the efficiency of the converter using $\eta = P_O/P_I$, where P_O is the average value of the output power and P_I is the average value of the input power, respectively.
6. This section may be repeated by plotting the power waveforms of all the components and then estimating their average values. All the power losses can be added to give the total power loss in the converter. Further, the efficiency can be estimated using $\eta = P_O/(P_{LS} + P_I)$, where P_O is the average value of the output power and P_{LS} is the sum of the average values of the power loss in individual components.

Post-lab Questions

1. Draw the waveforms of the currents through the magnetizing inductance and the output filter inductance. Describe the nature of currents through these components. Are the currents in these components in discontinuous-conduction mode?
2. Explain the operation of the forward converter in DCM.

24

Efficiency and DC Voltage Transfer Function of PWM Forward DC–DC Converter in DCM

Objectives

The objectives of this lab are:

- To design a PWM forward dc–dc converter in discontinuous-conduction mode (DCM) using the design equations.
- To analyze the variations in efficiency of the lossy forward converter in DCM at different load resistances and different input voltages.
- To observe the dependence of the lossy dc voltage transfer function on duty cycle.

Theory

The steady-state dc voltage transfer function of a converter is

$$M_{VDC} = \frac{V_O}{V_I}. \tag{24.1}$$

In terms of the circuit parameters, the transfer function under ideal conditions can be expressed as

$$M_{VDC} = \frac{2}{n_1 \left(1 + \sqrt{1 + \dfrac{8 f_s L}{D^2 R_L}} \right)}, \tag{24.2}$$

Laboratory Manual for Pulse-Width Modulated DC–DC Power Converters, First Edition.
Marian K. Kazimierczuk and Agasthya Ayachit.
© 2016 John Wiley & Sons, Ltd. Published 2016 by John Wiley & Sons, Ltd.

resulting in

$$D = n_1 M_{VDC} \sqrt{\frac{2f_s L}{R_L(1 - n_1 M_{VDC})}}. \tag{24.3}$$

The overall converter efficiency is

$$\eta = \frac{1}{1 + \dfrac{P_{LS}}{P_O}}, \tag{24.4}$$

where P_{LS} is the total power loss in the converter in terms of M_{VDC} is

$$P_{LS} = \left[\frac{2}{3} \left(\frac{r_{DS} + r_{T1}}{n_1^2} + R_F + r_{T2} \right) n_1 M_{VDC} \sqrt{\frac{2(1 - n_1 M_{VDC})}{f_s L R_L}} + \frac{f_s C_o R_L}{M_{VDC}^2} \right.$$
$$\left. + \frac{2R_F}{3} \sqrt{\frac{2(1 - n_1 M_{VDC})^3}{f_s L R_L}} + \frac{V_F}{V_O} + \frac{2r_L}{3} \sqrt{\frac{2(1 - n_1 M_{VDC})}{f_s L R_L}} \right] P_O, \tag{24.5}$$

where r_L is the equivalent series resistance (ESR) of the output filter inductor, r_{DS} is the on-state resistance of the MOSFET, R_L is the load resistance, R_F is the forward resistance of the diode, r_C is the ESR of the output filter capacitor, V_F is the forward voltage drop of the diode, f_s is the switching frequency, and C_o is the output capacitance of the MOSFET, r_{T1} and r_{T2} are the ESR of the primary and secondary windings of the transformer, and n_1 is the turns ratio between the primary and secondary windings.

The steady-state dc voltage transfer function of the lossy forward converter in DCM is

$$M_{VDC} = \frac{2}{n_1 \left(1 + \sqrt{1 + \dfrac{8f_s L}{\eta D^2 R_L}} \right)}. \tag{24.6}$$

Specifications

The specifications of the buck converter are as provided in Table 24.1.

Pre-lab

For the specifications provided, find the values of all the components and specifications for the forward dc–dc converter operating in DCM using the relevant design equations provided in Tables B.2 and B.3 in Appendix B.

Quick Design

Assume an overall efficiency of $\eta = 80\%$ and $D_{max} = 0.4$. The turns ratio between primary and secondary windings $n_1 = 8$ and the turns ratio between the primary and tertiary (reset) windings $n_3 = 8$. Choose:

Table 24.1   Parameters and component values

Parameters	Notation	Value
Minimum rms input voltage	$V_{Imin(rms)}$	85 V
Minimum dc input voltage	V_{Imin}	120.21 V
Maximum rms input voltage	$V_{Imax(rms)}$	264 V
Maximum dc input voltage	V_{Imax}	373.35 V
DC output voltage	V_O	5 V
Switching frequency	f_s	100 kHz
Maximum output current	I_{Omax}	20 A
Minimum output current	I_{Omin}	0 A
Output voltage ripple	V_r	$< 0.01 V_O$

$L_m = 0.5$ mH, $L = 0.56\mu$H, ESR of the inductor L is $r_L = 0.015$ Ω, $R_{Lmin} = 0.25$ Ω, $D_{min} = 0.0849$, $D_{max} = 0.3055$ (or assumed value), $C = 1$ mF, ESR of the capacitor $r_C = 7$ mΩ, ESR of the primary and secondary windings are $r_{T1} = 50$ mΩ, $r_{T2} = 10$ mΩ, respectively.

MOSFET: International Rectifier IRF740 n-channel power MOSFET with $V_{DSS} = 400$ V, $I_{SM} = 10$ A, $r_{DS} = 0.55$ Ω at $T = 25°$C, $C_o = 100$ pF, and $V_t = 4$ V.

Diode: Two ON Semiconductor MBR2540 diodes for D_1 and D_2 with $V_{RRM} = 40$ V, $I_F = 25$ A, $R_F = 10$ mΩ, and $V_F = 0.3$ V and a MR826 fast recovery diode for D_3 with $V_{RRM} = 600$ V, $I_{F(AV)} = 35$ A. $R_F = 10$ mΩ, and $V_F = 1.2$ V.

Procedure

A. Efficiency of the Forward Converter as a Function of the Input Voltage at Different Load Conditions

1. The equations for the efficiency and total power loss of the lossy forward converter are given in Equations (24.4) and (24.5). In Equation (24.5), replace the term M_{VDC} with V_O/V_I. Initially, let $R_L = 2.5$ Ω.
2. Define the equation for the efficiency as given in Equations (20.4) and (20.5) on MATLAB® along with all the specifications and their values. Define an input voltage range, that is, vary V_I from 100 to 400 V in steps of 1 V.
3. Plot the efficiency as a function of the input voltage for $R_L = 75$ Ω.
4. Repeat the activity for $R_L = 0.25$ Ω and $R_L = 0.5$ Ω. Plot all the three curves on the same figure window. Use `legend` or `text` commands to label the different curves clearly.

B. Efficiency of the Forward Converter as a Function of the Output Current at Minimum, Nominal, and Maximum Input Voltages

1. Use the MATLAB® code developed in Section A. Let M_{VDC} be fixed and be equal to $V_I = V_{Imin} = 120.21$ V. Replace the term R_L with V_O/I_O.
2. Define the equation for the efficiency on MATLAB® along with all the specifications and their values. Define a range for the output current, that is, vary I_O from 0 to 2 A in steps of 0.001 A.

3. Plot the efficiency as a function of the output current for $V_I = V_{Imin}$.
4. Repeat the activity for $V_{Imax} = 373.35$ V. Plot the two curves on the same figure window. Use `legend` or `text` commands to label the different curves clearly.

C. DC Voltage Transfer Function of the Forward Converter as a Function of the Duty Cycle

1. The equation for the dc voltage transfer function of the lossy forward converter is given in Equation (24.6).
2. Define the equation for the dc voltage transfer function on MATLAB® along with all the specifications and their values. Define a duty cycle range, that is, vary D from 0 to 1 in steps of 0.001.
3. Plot the dc voltage transfer function M_{VDC} as a function of the duty cycle for $R_L = R_{Lmax}$.

Post-lab Question

Explain the significance of all the plots obtained in Sections A, B, and C.

25

Half-Bridge DC–DC Converter in CCM—Steady-State Simulation

Objectives

The objectives of this lab are:

- To design a pulse-width modulated half-bridge dc–dc converter operating in continuous-conduction mode (CCM) for the design specifications provided.
- To simulate the converter and analyze its steady-state characteristics.
- To estimate the overall efficiency of the half-bridge converter.

Specifications

The specifications of the half-bridge converter are as provided in Table 25.1.

Pre-lab

For the above specifications, find the values of all the components and parameters for the half-bridge dc–dc converter operating in CCM using the relevant design equations.

Quick Design

Assume an overall efficiency of $\eta = 75\%$. The turns ratio between primary and secondary windings $n = 7$. Choose:

$L_m = 500$ µH, $L = 20$ µH, ESR of the inductor L is $r_L = 0.01$ Ω, $R_{Lmin} = 0.125$ Ω, $R_{Lmax} = 1.25$ Ω, $D_{min} = 0.2496$, $D_{nom} = 0.299$, $D_{max} = 0.3674$, $C = 47$ µF, ESR of the capacitor $r_C = 50$ mΩ, ESR of the primary and secondary windings are $r_{T1} = 20$ mΩ, $r_{T2} = r_{T3} = 5$ mΩ.

Laboratory Manual for Pulse-Width Modulated DC–DC Power Converters, First Edition.
Marian K. Kazimierczuk and Agasthya Ayachit.
© 2016 John Wiley & Sons, Ltd. Published 2016 by John Wiley & Sons, Ltd.

Table 25.1   Parameters and their values

Parameter	Notation	Value
Minimum rms input voltage	$V_{Imin(rms)}$	90 V
Minimum dc input voltage	V_{Imin}	127 V
Nominal rms input voltage	$V_{Inom(rms)}$	110 V
Nominal dc input voltage	V_{Inom}	156 V
Maximum rms input voltage	$V_{Imax(rms)}$	132 V
Maximum dc input voltage	V_{Imax}	187 V
DC output voltage	V_O	5 V
Switching frequency	f_s	100 kHz
Maximum output current	I_{Omax}	40 A
Minimum output current	I_{Omin}	4 A
Output voltage ripple	V_r	$< 0.01 V_O$

MOSFET: International Rectifier IRF640 n-channel power MOSFET with $V_{DSS} = 200$ V, $I_{SM} = 18$ A, $r_{DS} = 180$ mΩ at $T = 25°$C, $C_o = 100$ pF, and $V_t = 4$ V.

Diode: Two ON Semiconductor MBR2545CT diodes for D_1 and D_2 with $V_{RRM} = 45$ V, $I_{F(AV)} = 30$ A, $I_{FM} = 300$ A, $R_F = 13.25$ mΩ, and $V_F = 0.27$ V.

Procedure

A. Analysis of the Half-Bridge Converter in Steady State

1. Construct the circuit of the half-bridge converter shown in Figure 25.1 on the circuit simulator. Refer to Lab 16 for the method to use a transformer. Choose an appropriate number of turns. You may select $N_1 = 14$ and $N_2 = N_3 = 2$.
2. Enter the values of all the components. Place a pulse voltage source in order to provide the gate-to-source voltage at the MOSFET terminals. Set `time period` = 10 μs, `duty cycle/width` = 0.2999, and `amplitude` = 12 V. Let the `rise time` and `fall time` be equal to zero (optional). Let the input voltage V_I be equal to $V_{Inom} = 156$ V.

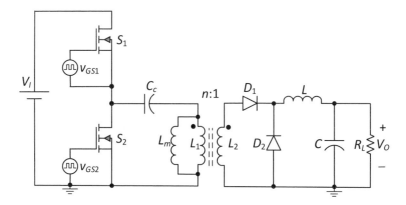

Figure 25.1   Circuit diagram of the PWM half-bridge dc–dc converter with half-wave rectifier.

3. Set simulation type to `transient analysis`. Set `end time` = 20 ms and `time step` = 0.1 µs. Run the simulation.
4. Plot the following parameters after successful completion of the simulation. You may display the waveforms on different figure windows for better clarity.
   - Gate-to-source voltages v_{GS1}, v_{GS2} of MOSFETs S_1 and S_2, drain-to-source voltages v_{DS1}, v_{DS2} of the MOSFETs S_1 and S_2.
   - Diode voltages v_{D1}, v_{D2} of the diodes D_1 and D_2.
   - Output voltage v_O, output current i_O, and output power p_O.
   - Current through the magnetizing inductance i_{Lm}, current through the output filter inductance i_L, diode currents i_{D1}, i_{D2}.
   - Currents i_{S1} and i_{S2} of the MOSFETs S_1 and S_2.
   Use the `zoom` option to focus only on the steady-state region.
5. Observe the waveform of the current through the output filter inductance to ensure whether it is in CCM. If the current is not in CCM, then increase the value of the inductance L and repeat the simulation.
6. For the above-mentioned waveforms, measure:
   - The average and peak-to-peak values of the current through the magnetizing inductance and the output filter inductance.
   - The maximum and average values of the voltage across the MOSFETs.
   - The minimum and average values of the voltage across the diodes D_1, and D_2.
   - The maximum and average values of the currents through the MOSFETs and the diodes.
   Ensure that the values obtained above match the desired specifications.
7. Execute the above steps by changing the duty cycle to $D_{max} = 0.3674$ and $V_{Imin} = 127$ V. Repeat for $D_{min} = 0.2496$ and $V_{Imax} = 187$ V.

B. Estimation of Power Losses and Overall Efficiency

1. Plot the waveforms of the input power p_I and the output power p_O at $R_L = R_{Lmin}$. Zoom in to the steady-state region.
2. Measure the average values of the input power and the output power. Consider the magnitude of the average value of the input power. Calculate the efficiency using $\eta = P_O/P_I$, where P_O is the average value of the output power and P_I is the average value of the input power, respectively.
3. This section may be repeated by plotting the power waveforms of all the components and then estimating their average values. All the power losses can be added to give the total power loss in the converter. Further, the efficiency can be estimated using $\eta = P_O/(P_{LS} + P_I)$, where P_O is the average value of the output power and P_{LS} is the sum of the average values of the power loss in individual components.

Post-lab Questions

1. What is the purpose of the coupling capacitor C_c on the primary-side of the half-bridge converter?

2. What is the maximum achievable duty cycle for the transistors in the half-bridge switching network?
3. What are the maximum current and voltage stresses of the MOSFETs in the half-bridge switching network?
4. Draw the circuit of the half-bridge switching network with a full-wave rectifier. What are the advantages of using full-wave rectification?

26

Efficiency and DC Voltage Transfer Function of PWM Half-Bridge DC–DC Converter in CCM

Objectives

The objectives of this lab are:

- To design a PWM half-bridge dc–dc converter in continuous-conduction mode (CCM) using the design equations.
- To analyze the variations in efficiency of the lossy half-bridge converter in CCM at different load resistances and different input voltages.
- To observe the dependence of the lossy dc voltage transfer function on duty cycle.

Theory

The steady-state dc voltage transfer function of a converter is

$$M_{VDC} = \frac{V_O}{V_I}. \tag{26.1}$$

The overall converter efficiency is

$$\eta = \frac{N_\eta}{D_\eta}, \tag{26.2}$$

Laboratory Manual for Pulse-Width Modulated DC–DC Power Converters, First Edition.
Marian K. Kazimierczuk and Agasthya Ayachit.
© 2016 John Wiley & Sons, Ltd. Published 2016 by John Wiley & Sons, Ltd.

where

$$N_\eta = 1 - \frac{M_{VDC}[2(r_{DS} + r_{T1} + r_{Cb})]}{nR_L} - \frac{nM_{VDC}(R_F + r_{T2})}{R_L} + \frac{r_C R_L n M_{VDC}}{12 f_s^2 L^2}$$

$$+ \left\{ \left[\frac{M_{VDC}[2(r_{DS} + r_{T1} + r_{Cb})]}{nR_L} + \frac{nM_{VDC}(R_F + r_{T2})}{R_L} - \frac{r_C R_L n M_{VDC}}{12 f_s^2 L^2} - 1 \right]^2 \right.$$

$$\left. - \frac{r_C R_L n^2 M_{VDC}^2}{3 f_s^2 L^2} \left[1 + \frac{r_L}{R_L} + \frac{R_F + r_{T2}}{2R_L} + \frac{V_F}{V_O} + \frac{2 f_s C_o R_L}{M_{VDC}^2 + \frac{r_C R_L}{48 f_s^2 L^2}} \right] \right\}^{\frac{1}{2}}, \qquad (26.3)$$

and

$$D_\eta = 2 \left(1 + \frac{r_L}{R_L} + \frac{R_F + r_{T2}}{2R_L} + \frac{V_F}{V_O} + \frac{2 f_s C_o R_L}{M_{VDC}^2 + \frac{r_C R_L}{48 f_s^2 L^2}} \right), \qquad (26.4)$$

where r_L is the equivalent series resistance (ESR) of the output filter inductor, r_{DS} is the on-state resistance of the MOSFET, R_L is the load resistance, R_F is the forward resistance of the diode, r_C is the ESR of the output filter capacitor, V_F is the forward voltage drop of the diode, f_s is the switching frequency, and C_o is the output capacitance of the MOSFET, r_{T1} and r_{T2} are the ESR of the primary and secondary windings of the transformer, and n is the turns ratio between the primary and secondary windings.

The steady-state dc voltage transfer function of the lossy half-bridge converter is

$$M_{VDC} = \frac{\eta D}{n}. \qquad (26.5)$$

Specifications

The specifications of the buck converter are as provided in Table 26.1

Pre-lab

For the above specifications, find the values of all the components and parameters of the half-bridge dc–dc converter operating in CCM using the relevant design equations.

Quick Design

Assume an overall efficiency of $\eta = 75\%$. The turns ratio between primary and secondary windings $n = 7$. Choose:

Table 26.1   Parameters and their values

Parameter	Notation	Value
Minimum rms input voltage	$V_{Imin(rms)}$	90 V
Minimum dc input voltage	V_{Imin}	127 V
Nominal rms input voltage	$V_{Inom(rms)}$	110 V
Nominal dc input voltage	V_{Inom}	156 V
Maximum rms input voltage	$V_{Imax(rms)}$	132 V
Maximum dc input voltage	V_{Imax}	187 V
DC output voltage	V_O	5 V
Switching frequency	f_s	100 kHz
Maximum output current	I_{Omax}	40 A
Minimum output current	I_{Omin}	4 A
Output voltage ripple	V_r	$< 0.01 V_O$

$L_m = 500$ µH, $L = 20$ µH, ESR of the inductor L is $r_L = 0.01$ Ω, $R_{Lmin} = 0.125$ Ω, $R_{Lmax} = 1.25$ Ω, $D_{min} = 0.2496$, $D_{nom} = 0.299$, $D_{max} = 0.3674$ (or assumed value), $C = 47$ µF, ESR of the capacitor $r_C = 50$ mΩ, ESR of the primary and secondary windings are $r_{T1} = 20$ mΩ, $r_{T2} = r_{T3} = 5$ mΩ, $r_{Cb} = 50$ mΩ.

MOSFET: International Rectifier IRF640 n-channel power MOSFET with $V_{DSS} = 200$ V, $I_{SM} = 18$ A, $r_{DS} = 180$ mΩ at $T = 25°$ C, $C_o = 100$ pF, and $V_t = 4$ V.

Diode: Two ON Semiconductor MBR2545CT diodes for D_1 and D_2 with $V_{RRM} = 45$ V, $I_{F(AV)} = 30$ A, $I_{FM} = 300$ A, $R_F = 13.25$ mΩ, and $V_F = 0.27$ V.

Procedure

A. Efficiency of the Half-bridge Converter as a Function of the Input Voltage at Full and Light Load Conditions

1. The equation for the efficiency of the lossy half-bridge converter is given in Equation (26.2). Replace the term M_{VDC} in Equations (26.2), (26.3), and (26.4) with V_O/V_I. Initially, let $R_L = R_{Lmin} = V_O/I_{Omax}$.
2. Define the equation for the efficiency on MATLAB® along with all the specifications and their values. Define an input voltage range, that is, vary V_I from 90 to 220 V in steps of 0.01 V.
3. Plot the efficiency as a function of the input voltage $R_L = R_{Lmin}$.
4. Repeat the activity by replacing R_L with $R_{Lmax} = V_O/I_{Omin}$. Plot the two curves on the same figure window. Use legend or text commands to label the different curves clearly.

B. Efficiency of the Half-bridge Converter as a Function of the Output Current at Minimum, Nominal, and Maximum Input Voltages

1. Use the code developed in Section A. In this activity, let the value of M_{VDC} be fixed, that is, let $V_I = V_{Imin}$. Also, replace the term R_L with V_O/I_O.

2. Vary I_O from 1 to 10 A in steps of 0.001 A.
3. Plot the efficiency as a function of the output current for $V_I = V_{Imin}$.
4. Repeat the activity for V_{Inom} and V_{Imax}. Plot all the three curves on the same figure window. Use `legend` or `text` commands to label the different curves clearly.

C. DC Voltage Transfer Function of the Half-bridge Converter as a Function of the Duty Cycle

1. In the expression for the efficiency in Equations (26.2), (26.3), and (26.4), replace M_{VDC} with D/n.
2. The equation for the dc voltage transfer function of the lossy half-bridge converter is given in Equation (26.5). Define the equation for the dc voltage transfer function on MATLAB® along with all the specifications and their values. Replace R_L with $R_{Lmin} = V_O/I_{Omax}$.
3. Define a duty cycle range, that is, vary D from 0 to 1 in steps of 0.001.
4. Plot the dc voltage transfer function M_{VDC} as a function of the duty cycle for $R_L = R_{Lmin}$.
5. Repeat the activity using $R_L = R_{Lmax}$. Plot the two curves on the same figure window. Use `legend` or `text` commands to label the different curves clearly.

Post-lab Question

Explain the significance of all the plots obtained in Sections A, B, and C.

27

Full-Bridge DC–DC Converter in CCM—Steady-State Simulation

Objectives

The objectives of this lab are:

- To design a pulse-width modulated full-bridge dc–dc converter operating in continuous-conduction mode (CCM) for the design specifications provided.
- To analyze the characteristics of the converter in steady state.
- To estimate the losses and overall efficiency of the full-bridge converter.

Specifications

The specifications of the full-bridge converter are as provided in Table 27.1.

Pre-lab

For the above specifications, find the values of all the components and parameters for the forward dc–dc converter operating in CCM using the relevant design equations.

Quick Design

Assume an overall efficiency of $\eta = 85\%$ and $D_{max} = 0.4$. The turns ratio between primary and secondary windings $n = 4$. Choose:

$L_m = 3.5$ mH, $L = 40$ μH, ESR of the inductor L is $r_L = 0.01$ Ω, $R_{Lmin} = 1.92$ Ω, $R_{Lmax} = 19.2$ Ω, $D_{min} = 0.3322$, $D_{nom} = 0.3631$, $D_{max} = 0.3991$ (or assumed value), $C = 50$ μF, ESR of the capacitor $r_C = 100$ mΩ, ESR of the primary and secondary windings are $r_{T1} = 0.025$ Ω, $r_{T2} = r_{T3} = 0.01$ Ω.

Laboratory Manual for Pulse-Width Modulated DC–DC Power Converters, First Edition.
Marian K. Kazimierczuk and Agasthya Ayachit.
© 2016 John Wiley & Sons, Ltd. Published 2016 by John Wiley & Sons, Ltd.

Table 27.1   Parameters and component values

Parameters	Notation	Value
Minimum rms input voltage	$V_{Imin(rms)}$	200 V
Minimum dc input voltage	V_{Imin}	283 V
Nominal rms input voltage	$V_{Inom(rms)}$	220 V
Nominal dc input voltage	V_{Inom}	311 V
Maximum rms input voltage	$V_{Imax(rms)}$	240 V
Maximum dc input voltage	V_{Imax}	340 V
DC output voltage	V_O	48 V
Switching frequency	f_s	100 kHz
Maximum output current	I_{Omax}	25 A
Minimum output current	I_{Omin}	2.5 A
Output voltage ripple	V_r	$< 0.01 V_O$

MOSFET: International Rectifier IRF740 n-channel power MOSFET with $V_{DSS} = 400$ V, $I_{SM} = 10$ A, $r_{DS} = 550$ mΩ at $T = 25°$ C, $C_o = 100$ pF, and $V_t = 4$ V.

Diode: Two American Semiconductor MR866 diodes for D_1 and D_2 with $V_{RRM} = 600$ V, $I_{F(AV)} = 40$ A, $I_{FM} = 300$ A, $R_F = 12.5$ mΩ, and $V_F = 0.7$ V.

Procedure

A. Simulation of the Full-Bridge Converter

1. Construct the circuit of the half-bridge converter shown in Figure 27.1 on the circuit simulator. You may use (a) a dc/dc ideal transformer and place an inductor L_m in parallel or (b) use the model of the two-winding transformer as shown at the end of this lab. Choose an appropriate number of turns. You may select $N_1 = 8$ and $N_2 = N_3 = 2$.
2. Initially, assign the input voltage as $V_{Inom} = 311$ V at a duty cycle of $D_{nom} = 0.3631$. Let the value of the load resistor be $R_L = R_{Lmin} = 1.92$ Ω.
3. Enter the values of all the components. Connect a pulse voltage source between the gate and source terminals of the MOSFET. Set `time period = 10` μs, `duty cycle/`

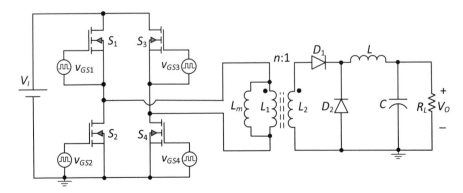

Figure 27.1   Circuit diagram of the PWM full-bridge dc–dc converter with half-wave rectifier.

width = 0.3631, and amplitude = 12 V. Let the rise time and fall time be equal to zero (optional).

4. Set simulation type to transient analysis. Set end time = 20 ms and a time step = 0.1 µs. Run the simulation.
5. Plot the following parameters after successful completion of the simulation. You may display the waveforms on different figure windows for better clarity.
   - Gate-to-source voltages v_{GS1}, v_{GS2}, v_{GS3}, v_{GS4} of MOSFETs S_1, S_2, S_3 and S_4, drain-to-source voltages v_{DS1}, v_{DS2}, v_{DS3}, v_{DS4} of the MOSFETs S_1, S_2, S_3 and S_4.
   - Diode voltages v_{D1}, v_{D2} of the diodes D_1 and D_2.
   - Output voltage V_O, output current I_O, and output power P_O.
   - Current through the magnetizing inductance i_{Lm}, current through the output filter inductance i_L, diode currents i_{D1}, i_{D2}.
   - Currents i_{S1}, i_{S2}, i_{S3}, and i_{S4} of the MOSFETs S_1, S_2, S_3, and S_4.
   Use the zoom option to display only the steady-state region.
6. Observe the waveform of the current through the output filter inductance to ensure whether it is in CCM. If the current is not in CCM, then increase the value of the inductance L and repeat the simulation.
7. For the above-mentioned waveforms, measure:
   - The average and peak-to-peak values of the current through the magnetizing inductance and the output filter inductance.
   - The maximum and average values of the voltage across all the MOSFETs.
   - The minimum and average values of the voltage across all the diodes.
   - The maximum and average values of the currents through the MOSFETs and the diodes.
   Ensure that the values obtained above match the desired specifications.
8. Execute the above steps by changing the duty cycle to $D_{max} = 0.3991$ and input voltage to $V_{Imin} = 283$ V. Repeat for $D_{min} = 0.3322$ and $V_{Imax} = 340$ V.

B. Estimation of the Overall Efficiency

1. Set up the converter to operate at nominal operating condition, that is, at $D = D_{nom}$ and $V_I = V_{Inom}$. Let $R_L = R_{Lmin}$ such that the converter delivers the maximum output power.
2. Set the simulation type to transient analysis and perform the simulation.
3. Plot the waveforms of the input power p_I and the output power p_O. Zoom in to the steady-state region.
4. Measure the average values of the input power and the output power. If the input power is negative, then consider only the magnitude of the average value of the input power.
5. Calculate the efficiency of the converter using $\eta = P_O/P_I$, where P_O is the average value of the output power and P_I is the average value of the input power, respectively.
6. This section may be repeated by plotting the power waveforms of all the components and then estimating their average values. All the power losses can be added to give the total power loss in the converter. Further, the efficiency can be estimated using $\eta = P_O/(P_{LS} + P_I)$, where P_O is the average value of the output power and P_{LS} is the sum of the average values of the power loss in individual components.

Post-lab Questions

1. Give the expression for the dc voltage transfer function of the lossless full-bridge converter.
2. What is the maximum value of the duty cycle for the full-bridge converter?
3. How can cross-conduction be prevented in the full-bridge converter?
4. Is the magnetizing inductance in the full-bridge converter needed to store energy?

Note

An example model of a two-winding transformer developed using PSpice can be obtained from Lab 16. The circuit can be modified into a three-winding transformer by adding a tertiary winding and including the component name in the code for coefficient coupling.

28

Efficiency and DC Voltage Transfer Function of PWM Full-Bridge DC–DC Converters in CCM

Objectives

The objectives of this lab are:

- To design a PWM full-bridge dc–dc converter in continuous-conduction mode (CCM) using the design equations.
- To analyze the variations in efficiency of the lossy full-bridge converter in CCM at different load resistances and different input voltages.
- To observe the dependence of the lossy dc voltage transfer function on duty cycle.

Theory

The steady-state dc voltage transfer function of a converter is

$$M_{VDC} = \frac{V_O}{V_I}. \tag{28.1}$$

The overall converter efficiency is

$$\eta = \frac{N_\eta}{D_\eta}, \tag{28.2}$$

Laboratory Manual for Pulse-Width Modulated DC–DC Power Converters, First Edition.
Marian K. Kazimierczuk and Agasthya Ayachit.
© 2016 John Wiley & Sons, Ltd. Published 2016 by John Wiley & Sons, Ltd.

Table 28.1　Parameters and component values

Parameters	Notation	Value
Minimum rms input voltage	$V_{Imin(rms)}$	90 V
Minimum dc input voltage	V_{Imin}	127 V
Nominal rms input voltage	$V_{Inom(rms)}$	110 V
Nominal dc input voltage	V_{Inom}	156 V
Maximum rms input voltage	$V_{Imax(rms)}$	132 V
Maximum dc input voltage	V_{Imax}	187 V
DC output voltage	V_O	5 V
Switching frequency	f_s	100 kHz
Maximum output current	I_{Omax}	40 A
Minimum output current	I_{Omin}	4 A
Output voltage ripple	V_r	$< 0.01 V_O$

where

$$N_\eta = 1 - \frac{nM_{VDC}}{2R_L}\left(\frac{4r_{DS} + r_{T1} + r_{Cc}}{n^2} + R_F + r_{T2}\right) + \frac{r_C R_L n M_{VDC}}{24 f_s^2 L^2}$$

$$+ \left\{\left[\frac{nM_{VDC}}{2R_L}\left(R_F + r_{T2} + \frac{4r_{DS} + 2r_{T1} + 2r_{Cb}}{n^2}\right) - \frac{r_C R_L n M_{VDC}}{24 f_s^2 L^2} - 1\right]^2 \right.$$

$$\left. - \frac{r_C R_L n^2 M_{VDC}^2}{12 f_s^2 L^2}\left[1 + \frac{r_L}{R_L} + \frac{R_F + r_{T2}}{2R_L} + \frac{V_F}{V_O} + \frac{4 f_s C_o R_L}{M_{VDC}^2 + \dfrac{r_C R_L}{48 f_s^2 L^2}}\right]\right\}^{\frac{1}{2}}, \qquad (28.3)$$

and

$$D_\eta = 2\left(1 + \frac{r_L}{R_L} + \frac{R_F + r_{T2}}{2R_L} + \frac{V_F}{V_O} + \frac{4 f_s C_o R_L}{M_{VDC}^2 + \dfrac{r_C R_L}{48 f_s^2 L^2}}\right), \qquad (28.4)$$

where r_L is the equivalent series resistance (ESR) of the output filter inductor, r_{DS} is the on-state resistance of the MOSFET, R_L is the load resistance, R_F is the forward resistance of the diode, r_C is the ESR of the output filter capacitor, V_F is the forward voltage drop of the diode, f_s is the switching frequency, and C_o is the output capacitance of the MOSFET, r_{T1} and r_{T2} are the ESR of the primary and secondary windings of the transformer, and n is the turns ratio between the primary and secondary windings.

The steady-state dc voltage transfer function of the lossy full-bridge converter is

$$M_{VDC} = \frac{2\eta D}{n}. \tag{28.5}$$

Specifications

The specifications of the buck converter are as provided in Table 26.1.

Pre-lab

For the above specifications, find the values of all the components and parameters for the forward dc–dc converter operating in CCM using the relevant design equations.

Quick Design

Assume an overall efficiency of $\eta = 75\%$. The turns ratio between primary and secondary windings $n = 7$. Choose:

$L_m = 500$ μH, $L = 20$ μH, ESR of the inductor L is $r_L = 0.01$ Ω, $R_{Lmin} = 0.125$ Ω, $R_{Lmax} = 1.25$ Ω, $D_{min} = 0.2496$, $D_{nom} = 0.299$, $D_{max} = 0.3674$ (or assumed value), $C = 47$ μF, ESR of the capacitor $r_C = 50$ mΩ, ESR of the primary and secondary windings are $r_{T1} = 20$ mΩ, $r_{T2} = r_{T3} = 5$ mΩ, $r_{Cb} = 50$ mΩ.

MOSFET: International Rectifier IRF640 n-channel power MOSFET with $V_{DSS} = 200$ V, $I_{SM} = 18$ A, $r_{DS} = 180$ mΩ at $T = 25°$ C, $C_o = 100$ pF, and $V_t = 4$ V.

Diode: Two ON Semiconductor MBR2545CT diodes for D_1 and D_2 with $V_{RRM} = 45$ V, $I_{F(AV)} = 30$ A, $I_{FM} = 300$ A, $R_F = 13.25$ mΩ, and $V_F = 0.27$ V.

Procedure

A. Efficiency of the Full-bridge Converter as a Function of the Input Voltage at Full and Light Load Conditions

1. The equation for the efficiency of the lossy full-bridge converter is given in Equation (28.2). Replace the term M_{VDC} in Equations (28.2), (28.3), and (28.4) with V_O/V_I. Initially, let $R_L = R_{Lmin} = V_O/I_{Omax}$.
2. Define the equation for the efficiency on MATLAB® along with all the specifications and their values. Define an input voltage range, that is, vary V_I from 280 to 340 V in steps of 0.01 V.
3. Plot the efficiency as a function of the input voltage $R_L = R_{Lmin}$.
4. Repeat the activity by replacing R_L with $R_{Lmax} = V_O/I_{Omin}$. Plot the two curves on the same figure window. Use legend or text commands to label the different curves clearly.

B. Efficiency of the Full-bridge Converter as a Function of the Output Current at Minimum, Nominal, and Maximum Input Voltages

1. Use the code developed in Section A. In this activity, let the value of M_{VDC} be fixed, that is, let $V_I = V_{Imin}$. Also, replace the term R_L with V_O/I_O.

2. Vary I_O from 0 to 40 A in steps of 0.001 A.
3. Plot the efficiency as a function of the output current for $V_I = V_{Imin}$.
4. Repeat the activity for V_{Inom} and V_{Imax}. Plot all the three curves on the same figure window. Use `legend` or `text` commands to label the different curves clearly.

C. DC Voltage Transfer Function of the Full-bridge Converter as a Function of the Duty Cycle

1. In the expression for efficiency in Equations (28.2), (28.3), and (28.4), replace M_{VDC} with $2D/n$.
2. The equation for the dc voltage transfer function of the lossy full-bridge converter is given in Equation (28.5). Define the equation for the dc voltage transfer function on MATLAB® along with all the specifications and their values. Replace R_L with $R_{Lmin} = V_O/I_{Omax}$.
3. Define a duty cycle range, that is, vary D from 0 to 1 in steps of 0.001.
4. Plot the dc voltage transfer function M_{VDC} as a function of the duty cycle for $R_L = R_{Lmin}$.
5. Repeat the activity using $R_L = R_{Lmax}$. Plot the two curves on the same figure window. Use `legend` or `text` commands to label the different curves clearly.

Post-lab Question

Explain the significance of all the plots obtained in Sections A, B, and C.

Part II

Closed-Loop Pulse-Width Modulated DC–DC Converters—Transient Analysis, Small-Signal Modeling, and Control

29

Design of the Pulse-Width Modulator and the PWM Boost DC–DC Converter in CCM

Objectives

The objectives of this lab are:

- To design a pulse-width modulator (op-amp comparator) needed to provide the gate-to-source voltage to the MOSFET.
- To design a PWM boost converter for the given specifications and simulate the converter with the pulse-width modulator.

Specifications

The specifications of the boost converter are as given in Table 29.1.

Pre-lab

For the specifications provided, find the values of all the components and specifications for the boost dc–dc converter operating in CCM using the relevant design equations provided in Table A.1 in Appendix A.

Quick Design

Assume an efficiency of 90%. Choose:

$L = 156\ \mu\text{H}$, ESR of the inductor $r_L = 0.19\ \Omega$, $R_{Lmin} = 40\ \Omega$, $R_{Lmax} = 160\ \Omega$, $D_{min} = 0.37$, $D_{nom} = 0.46$, $D_{max} = 0.55$, $C = 6.8\ \mu\text{F}$, ESR of the capacitor $r_C = 0.111\ \Omega$.

Laboratory Manual for Pulse-Width Modulated DC–DC Power Converters, First Edition.
Marian K. Kazimierczuk and Agasthya Ayachit.
© 2016 John Wiley & Sons, Ltd. Published 2016 by John Wiley & Sons, Ltd.

Table 29.1 Parameters of the boost converter

Parameter	Notation	Value
Minimum dc input voltage	V_{Imin}	10 V
Nominal dc input voltage	V_{Inom}	12 V
Maximum dc input voltage	V_{Imax}	14 V
DC output voltage	V_O	20 V
Switching frequency	f_s	100 kHz
Maximum output current	I_{Omax}	0.5 A
Nominal output current	I_{Onom}	0.3125 A
Minimum output current	I_{Omin}	0.125 A
Output voltage ripple	V_r	$< 0.01 V_O$

MOSFET: International Rectifier IRF142 n-channel power MOSFET with $V_{DSS} = 100$ V, $I_{SM} = 24$ A, $r_{DS} = 0.11\ \Omega$ at $T = 25^o$C, $C_o = 100$ pF, and $V_t = 4$ V.

Diode: ON Semiconductor MBR10100 with $V_{RRM} = 100$ V, $I_F = 10$ A, $R_F = 15\text{m}\Omega$, and $V_F = 0.8$ V.

Operational amplifier: LM741 or any ideal op-amp with provision for power supply terminals. Table 29.2 provides the values for the voltages and frequency related to the pulse-width modulator.

Procedure

A. Simulation of the Boost Converter Driven by the Pulse-Width Modulator

1. Construct the circuit shown in Figure 29.1 on the circuit simulator.
2. Provide a sawtooth-wave voltage at the inverting terminal of the op-amp. Provide the following details in order to generate the sawtooth voltage waveform: `initial voltage = 0 V`, `final/maximum voltage = 4 V`, `period = 10` μs, `rise time = 9.9` μs, `fall time = 0.1` μs, and `ON time/width/duty cycle = 0.1 ns`.
3. Provide a dc voltage at the noninverting terminal of the op-amp. The dc voltage behaves as the control voltage for the comparator/modulator. Initially, set the value of the dc voltage source to $V_{Cmin} = 1.48$ V and the input voltage to the boost converter at $V_I = V_{Imax} = 14$ V.
4. Set simulation type to `transient analysis`. Set `end time = 10 ms` and a `time step = 0.1` μs. Run the simulation.

Table 29.2 Parameters of the pulse-width modulator

Parameter	Notation	Value
Power supply voltage	$V_{CC}, -V_{EE}$	12 V, −12 V
Maximum amplitude of the triangular wave	V_{Tm}	4 V
Frequency of carrier wave	f_s	100 kHz
Minimum value of control voltage (for D_{min})	V_{Cmin}	1.48 V
Nominal value of control voltage (for D_{nom})	V_{Cnom}	1.84 V
Maximum value of control voltage (for D_{max})	V_{Cmax}	2.2 V

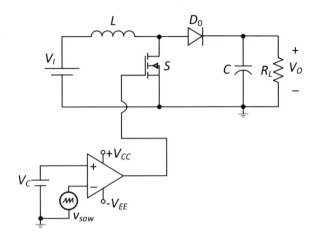

Figure 29.1   Circuit diagram of the PWM boost dc–dc converter with the pulse-width modulator.

5. Plot the following parameters after successful completion of the simulation. You may display the waveforms on different figure windows for better clarity.
   - Gate-to-source voltage v_{GS}, sawtooth voltage v_{saw}, and control voltage V_C.
   - Output voltage v_O, output current i_O, and output power p_O.
   - Inductor current i_L.
6. Observe the inductor current waveform to ensure whether the current is in CCM. If the current is not in CCM, then increase the value of the inductor and repeat the simulation.
7. Repeat the above steps by changing V_C to $V_{Cmax} = 2.2$ V and V_I to V_{Imin}. Similarly, redo the steps by changing V_C to $V_{Cnom} = 1.84$ V and V_I to V_{Inom}. Make sure the output voltage is equal to $V_O = 20$ V. If not, then adjust the value of V_C accordingly.

Post-lab Questions

1. Explain the principle of operation of the pulse-width modulator.
2. Explain the differences between using a sawtooth voltage waveform and a triangular voltage waveform at the input of the pulse-width modulator.
3. Describe the nature of the output voltage waveform, if the control voltage is sinusoidal.
4. For the specifications provided, determine the range of values of V_C for a variation in the duty cycle from 0 to 1.
5. How does the bandwidth of the op-amp affect the performance of the pulse-width modulator?

30

Dynamic Analysis of the Open-Loop PWM Boost DC–DC Converter in CCM for Step Change in the Input Voltage, Load Resistance, and Duty Cycle

Objective

The objective of this lab is to observe the response of the boost converter operating in continuous-conduction mode (CCM) for step changes in the input voltage, load resistance, and duty cycle.

Specifications

Use the data provided in Lab 29 for specifications of the boost converter and the pulse-width modulator.

Pre-lab

For the specifications provided, find the values of all the components and specifications for the boost dc–dc converter operating in CCM using the relevant design equations provided in Table A.1 in Appendix A.

Quick Design

Use the data from Lab 29 for the values of all the components.

Laboratory Manual for Pulse-Width Modulated DC–DC Power Converters, First Edition.
Marian K. Kazimierczuk and Agasthya Ayachit.
© 2016 John Wiley & Sons, Ltd. Published 2016 by John Wiley & Sons, Ltd.

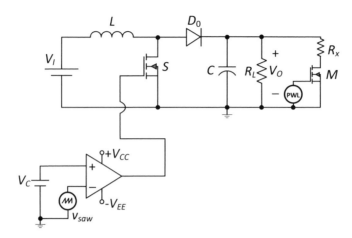

Figure 30.1 Circuit of the PWM boost dc–dc converter used to simulate step change in load resistance.

Procedure

A. Simulation of the Boost Converter for Step Change in the Load Resistance

1. Construct the circuit shown in Figure 30.1 on the circuit simulator. The circuit consists of a series connection of a resistor R_x and an ideal MOSFET M and is placed in parallel with the load resistor R_L. Let $R_x = 40\ \Omega$. This is needed in order to simulate the step change in the load resistance with respect to time.
2. Connect a piecewise-linear voltage source to the gate of the MOSFET M. Enter the details provided in Table 30.1 in order to provide a time-varying voltage at the gate of M.
3. Provide a sawtooth-wave voltage at the inverting terminal of the op-amp. Provide the following details in order to generate the sawtooth voltage waveform: `initial voltage = 0 V`, `final/maximum voltage = 4 V`, `period = 10 μs`, `rise time = 9.9 μs`, `fall time = 0.1 μs`, and `ON time/width/duty cycle = 0.1 ns`.
4. Connect a dc voltage at the noninverting terminal of the op-amp. The dc voltage behaves as the control voltage for the comparator/modulator. Initially, set the value of the dc voltage source to $V_{Cnom} = 1.84$ V and the input voltage to the boost converter at $V_I = V_{Inom} = 12$ V. Let the load resistance be $R_L = R_{Lmin} = 40\ \Omega$.
5. Set simulation type to `transient analysis`. Set `end time = 20 ms` and `time step = 0.1 μs`. Run the simulation.
6. Plot the following parameters after successful completion of the simulation. You may display the waveforms on different figure windows for better clarity.
   - Output voltage v_O, output current i_O, and output power p_O.
   - Inductor current i_L.
   Observe the transients which are generated at the instants of step change in the load resistance.

Table 30.1 Gate voltage profile of the MOSFET M

Time (ms)	0	5.99	6	13.9	14
Gate voltage V_G (V)	0	0	10	10	0

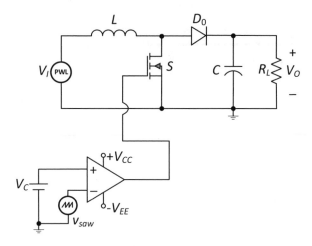

Figure 30.2 Circuit of the PWM boost dc–dc converter used to simulate step change in input voltage.

B. Simulation of the Boost Converter for Step Change in the Input Voltage

1. Construct the circuit shown in Figure 30.2 on the circuit simulator.
2. Remove the constant dc voltage supply at the input of the boost converter. Connect a piecewise-linear voltage source in its place and enter the details provided in Table 30.2 to provide a time-varying input voltage waveform.
3. The description of the sawtooth voltage waveform has been provided in Section A. Let the value of the control voltage be $V_C = V_{Cnom} = 1.84$ V. Let $R_L = R_{Lmin} = 40 \, \Omega$.
4. Set simulation type to `transient analysis`. Set end time = 20 ms and a time step = 0.1 μs. Run the simulation.
5. Plot the following parameters after successful completion of the simulation. You may display the waveforms on different figure windows for better clarity.
   - Input voltage v_I, output voltage v_O, output current i_O, and output power p_O.
   - Inductor current i_L.
   Observe the transients, which are generated at the instants of step change in the input voltage.

C. Simulation of the Boost Converter for Step Change in the Duty Cycle

1. Construct the circuit shown in Figure 30.3 on the circuit simulator.
2. Provide a piecewise-linear voltage source in place of the control voltage V_C at the input of the pulse-width modulator. Enter the details provided in Table 30.3 in order to provide a time-varying control voltage waveform:
3. The description of the sawtooth voltage waveform has been provided in Section A. Let the value of the input voltage be $V_I = V_{Inom} = 12$ V. Let $R_L = R_{Lmin} = 40 \, \Omega$.

Table 30.2 Input voltage profile for the PWL voltage source

Time (ms)		0	5.9	6	13.9	14	16.9
Input voltage V_I (V)	12	12		14	14	10	10

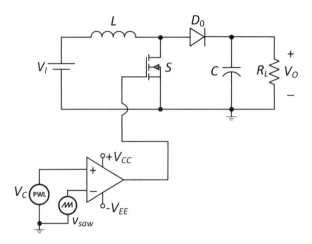

Figure 30.3 Circuit of the PWM boost dc–dc converter used to simulate step change in duty cycle.

Table 30.3 Control voltage profile for the PWL voltage source

Time (ms)	0	5.9	6	13.9	14	16.9	17
Control voltage V_C (V)	1.84	1.84	1.48	1.48	2.2	2.2	1.48

4. Set simulation type to `transient analysis`. Set `end time` = 20 ms and a `time step` = 0.1 μs. Run the simulation.
5. Plot the following parameters after successful completion of the simulation. You may display the waveforms on different figure windows for better clarity.
   - Output voltage v_O, output current i_O, and output power p_O.
   - Inductor current i_L.
   Observe the transients which are generated at the instants of step change in the duty cycle.

Post-lab Questions

1. In Section A, observe the waveform of the output voltage. Express your views on the transient characteristics such as undershoot, overshoot, and settling times of the output voltage waveform.
2. In Section B, observe the waveform of the output voltage. Discuss how the disturbance in the input voltage affects the nature of the output voltage waveform. Provide similar discussion for the results obtained in Section C.
3. Comment on the delay between the instant of change in the control voltage and the corresponding change in the output voltage.

31

Open-Loop Control-to-Output Voltage Transfer Function of the Boost Converter in CCM

Objectives

The objectives of this lab are:

- To understand the characteristics of the small-signal model of the boost converter in continuous-conduction mode (CCM).
- To obtain theoretically, the magnitude and phase plots of the control-to-output voltage transfer function T_p.
- To plot the response of the converter for step changes in the duty cycle using the control-to-output voltage transfer function.
- To simulate the small-signal model on a circuit simulator and compare the theoretical and simulation results.

PART A: BODE PLOTS AND STEP RESPONSE USING MATLAB®

Theory

The small-signal control-to-output voltage transfer function T_p of the boost dc–dc converter in CCM in terms of the impedances is given by

$$T_p = \frac{v_o}{d} = \frac{V_O}{1-D} \frac{1 - \frac{Z_1}{(1-D^2)R_L}}{1 + \frac{Z_1}{(1-D^2)Z_2}}, \tag{31.1}$$

Laboratory Manual for Pulse-Width Modulated DC–DC Power Converters, First Edition.
Marian K. Kazimierczuk and Agasthya Ayachit.
© 2016 John Wiley & Sons, Ltd. Published 2016 by John Wiley & Sons, Ltd.

where

$$Z_1 = r + sL,$$

$$Z_2 = \frac{R_L \left(r_C + \frac{1}{sC} \right)}{R_L + r_C + \frac{1}{sC}}. \tag{31.2}$$

Substituting Equation (31.2) into (31.1), we get

$$T_p = -\frac{V_O r_C}{(1-D)(R_L + r_C)} \frac{\left(s + \frac{1}{C r_C} \right) \left\{ s - \frac{1}{L}[R_L(1-D)^2 - r] \right\}}{s^2 + s \left\{ \frac{C[r(R_L + r_C) + (1-D)^2 R_L r_C] + L}{LC(R_L + r_C)} \right\} + \frac{r + (1-D)^2 R_L}{LC(R_L + r_C)}}. \tag{31.3}$$

In Equation (31.3), V_O is the output voltage of the converter, L is the inductance, C is the capacitance, R_L is the load resistance, D is the duty cycle, r_L is the equivalent series resistance (ESR) of the inductor, r_C is the ESR of the capacitor, and r is the equivalent averaged resistance given by

$$r = D r_{DS} + (1-D) R_F + r_L, \tag{31.4}$$

where r_{DS} is the on-state resistance of the MOSFET and R_F is the on-resistance of the diode.

Specifications

The specifications of the boost converter are as provided in Table 31.1.

Quick Design

Assume an efficiency of 90%. Choose:

$L = 156\ \mu H$, $r_L = 0.19\ \Omega$, $R_{Lmin} = 40\ \Omega$, $R_{Lnom} = 64\ \Omega$, $R_{Lmax} = 160\ \Omega$ $D = D_{nom} = 0.46$, $C = 6.8\ \mu F$, $r_C = 0.111\ \Omega$, $r_{DS} = 0.11\ \Omega$, $R_F = 15\ m\Omega$.

Table 31.1 Parameters of the boost converter

Parameter	Notation	Value
Minimum dc input voltage	V_{Imin}	10 V
Nominal dc input voltage	V_{Inom}	12 V
Maximum dc input voltage	V_{Imax}	14 V
DC output voltage	V_O	20 V
Switching frequency	f_s	100 kHz
Maximum output current	I_{Omax}	0.5 A
Nominal output current	I_{Onom}	0.3125 A
Minimum output current	I_{Omin}	0.125 A
Output voltage ripple	V_r	$< 0.01 V_O$

Procedure

A. *Magnitude and Phase Plot of the Control-to-Output Voltage Transfer Function*

1. Define all the parameters and variables on the MATLAB® editor window.
2. Define the $'s'$ function as s = tf('s'). This expression creates the variable $s = j\omega$ needed to define the transfer function in the Laplace domain.
3. Write the equation of the control-to-output voltage transfer function T_p in the impedance form given in Equation (31.1) or in the expanded form given in Equation (31.3).
4. Define the command bode(Tp), in order to plot the magnitude and phase of the control-to-output voltage transfer function.
5. On the figure, right-click to change the x-axis from ω-domain in rad/s to frequency domain in Hz.

Additional Activity 1: You may repeat the activities in Sections A for different values of load resistance. Adjust the value of the load resistance to $R_L = R_{Lmin} = 40\ \Omega$ and then to $R_L = R_{Lmax} = 160\ \Omega$. Variation in the load resistance causes a change in the value of the corner frequency of the boost converter, thereby affecting the location of the poles and zeros of the transfer function. (A for statement can be used for this purpose.)

B. *Response of the Boost Converter for Step Change in the Duty Cycle*

Write the following snippet of the code to plot the step response of the control-to-output voltage transfer function T_p.

```
t = 0:1e-7:3e-3;
u = step(Tp,t);
plot(t*1000,VO+(0.1*u))
grid on
xlabel('{\it t} (ms)')
ylabel('{\it v_o} (V)')
```

In the above code, the time axis is multiplied by 1000 in order to represent the x-axis in milliseconds. Record the values of the maximum overshoot, undershoot, rise time, settling time, and steady-state value after step change.

Additional Activity 2: Change the value of the duty cycle from $D_{nom} = 0.46$ to $D_{max} = 0.55$. Plot the response of the converter for step change in the duty cycle at the new duty cycle $D = 0.55$. Also, record the values of the maximum overshoot, undershoot, rise time, settling time, and steady-state value after step change. Repeat the activity for $D = 0.85$.

PART B: SIMULATION OF THE CIRCUIT MODEL

C. *Simulation of Small-Signal Model of the Boost Converter*

1. The small-signal model of the boost converter needed to determine the control-to-output voltage transfer function is as shown in Figure 31.1. Construct the model on the circuit

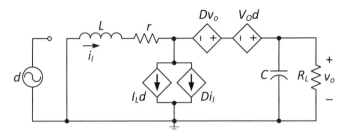

Figure 31.1 Small-signal model of the boost converter for determining the control-to-output voltage transfer function T_p with small-signal perturbations $v_i = 0$ and $i_o = 0$.

simulator. Make sure you have replaced the MOSFET with a parallel combination of current-controlled current source (for Di_l) and voltage-controlled current source (for $I_L d$). Similarly, the diode must be replaced with a series combination of two voltage-controlled voltage sources Dv_o and $V_O d$ connected in series.

2. Enter the values of the constants for all the dependent sources. To simulate the small-signal model at nominal operating conditions, let $D = D_{nom} = 0.46$, $R_L = R_{Lnom} = 64\ \Omega$, and $I_L = \frac{V_O}{R_L(1-D)} = 0.5787$ A.

3. Set the input voltage v_i to zero, that is, the small-signal ac excitation is only due to the duty cycle d.

4. Set the simulation type to `ac analysis`. Let the `start frequency` be 0, `end/stop frequency` be 1 MHz. Choose an appropriate step size depending on the resolution needed. (E.g., 1000 points per decade). Perform the simulation.

5. Plot the magnitude and phase of the output voltage v_o. Compare the theoretical plots in Section B, item 3 with those obtained through simulations.

Post-lab Questions

1. What is the significance of the control-to-output voltage transfer function?
2. Prepare a table listing the values of (a) poles, (b) zeros, and (c) the dc and low-frequency gain obtained using the transfer function presented in Equation (31.3).
3. Draw an asymptotic Bode plot for the control-to-output voltage transfer function using the zero-pole-gain data obtained in the above question.
4. Determine the following values from the magnitude plot obtained in Section A:
   - corner frequency f_0
   - 3-dB bandwidth
   - unity-gain crossover frequency
   - dc and low-frequency values of the transfer function
   - peak value of the values of the transfer function
5. From Additional Activity 1, tabulate the values of all the quantities mentioned in the previous question for the different load resistances.
6. Determine the expressions for the poles and zeros from the control-to-output voltage transfer function. Verify if the values of the poles and zeros obtained theoretically is in accordance with those obtained through MATLAB®.
7. In Section B, determine the values of the following quantities:
   - steady-state voltage value after step change
   - steady-state error

- rise time
- settling time
- maximum overshoot

8. From the plot for step response, observe the undershoot present in the output voltage waveform at $t = 0$. Why does an undershoot occur in the output voltage waveform for step change in the duty cycle? What is the effect on the value of the undershoot if the steady-state value of the duty cycle is increased?

9. From Additional Activity 2, what is the effect on the transfer function at a duty cycle of $D = 0.85$. Is the boost converter in open loop stable? What is the value of the steady state output voltage at this duty cycle?

10. Verify whether the plots obtained in Section C match with the results from Section A.

Note

1. *Changing the units on x-axis*: The units on the *x*-axis of the Bode plot is preset as radians per second. The units may be changed to Hertz using the following command:

```
set(cstprefs.tbxprefs,'FrequencyUnits','Hz');
```

2. *Method to draw the magnitude and phase plots by predefining a frequency variable.*
   - Define a variable f. Let f vary from 0 to 1 MHz. Define the term s such that $s = 2\pi f$.
   - Define the expression of the small-signal transfer function. Include a dot ($\cdot$) before every multiplication, division, or power symbol to enable element-wise operation. Make sure you include a dot in the definition of the term s also.
   - You may use the `plot` command. Let the *x*-axis be the frequency variable f. The *y*-axis must be the magnitude or the phase of the transfer function. In order to convert the magnitude of a transfer function *TF* from a linear scale to decibels, use `20*log10(TF)`. Similarly, to plot the phase in degrees, use `angle(TF)*180/pi`.

32

Root Locus and 3D Plot of the Control-to-Output Voltage Transfer Function

Objectives

The objectives of this lab are:

- To plot the root locus of the control-to-output voltage transfer function in order to determine the location and trajectory of the poles and zeros.
- To plot a 3D graph representing the dependence of the magnitude of the control-to-output voltage transfer function on the complex frequency.

Theory

The expression for the control-to-output voltage transfer function T_p in the standard second-order form is represented as

$$T_p = T_{p0} \frac{s^2 - s(\omega_{zn} + \omega_{zp}) + \omega_{zn}\omega_{zp}}{s^2 + 2\xi\omega_0 s + \omega_0^2}, \qquad (32.1)$$

where

$$T_{p0} = \frac{V_O r_C}{(1 - D)(R_L + r_C)},$$

$$\omega_{zn} = -\frac{1}{C r_C},$$

$$\omega_{zp} = \frac{R_L(1 - D)^2 - r}{L},$$

$$\xi = \frac{L + C[r(R_L + r_C) + (1 - D)^2 R_L r_C]}{2\sqrt{LC(R_L + r_C)[(1 - D)^2 R_L + r]}},$$

Laboratory Manual for Pulse-Width Modulated DC–DC Power Converters, First Edition.
Marian K. Kazimierczuk and Agasthya Ayachit.
© 2016 John Wiley & Sons, Ltd. Published 2016 by John Wiley & Sons, Ltd.

and

$$\omega_0 = \sqrt{\frac{(1-D)^2 R_L + r}{LC(R_L + r_C)}}.$$

In general, Equation (32.1) can be represented as

$$T_p = T_{p0} \frac{a_2 s^2 + a_1 s + a_0}{b_2 s^2 + b_1 s + b_0} \tag{32.2}$$

where

$$a_2 = 1, \qquad a_1 = -(\omega_{zn} + \omega_{zp}), \qquad a_0 = \omega_{zn}\omega_{zp},$$

and

$$b_2 = 1, \qquad b_1 = 2\xi\omega_o, \qquad b_0 = \omega_0^2.$$

Quick Design

Choose:

$L = 156$ μH, $r_L = 0.19$ Ω, $R_L = R_{Lnom} = 64$ Ω, $D = 0.46$, $C = 6.8$ μF, $r_C = 0.111$ Ω, $r_{DS} = 0.11$ Ω, $R_F = 15$ mΩ.

Procedure

A. Root Locus Plot of the Control-to-Output Voltage Transfer Function

1. Define all the parameters and variables on the MATLAB® editor window.
2. Define the $'s'$ function as s = tf('s'). This expression creates the variable $s = j\omega$ needed to define the transfer function in the Laplace domain.
3. The expression for the control-to-output voltage transfer function is given in Equation (32.1). Define the expression of the transfer function T_p on MATLAB® in the form given in Equation (32.2). Define the numerator and denominator as given below:

```
Tpnum = Tpo*[a2 a1 a0];
Tpden = [b2 b1 b0];
```

Further, define the transfer function T_p as

```
Tp = tf(Tpnum, Tpden);
```

4. Use command rlocus(Tp) to plot the root locus of the control-to-output voltage transfer function.
5. Determine the location of the poles and zeros on the root locus plot. Also, record the trajectory of the poles and zeros.
6. Change the duty cycle D to 0.85. Repeat the above steps.

B. 3D Plot for the Control-to-Output Voltage Transfer Function

1. Use the terms Tpnum and Tpden defined in step 2 in Section A. Use $D = 0.46$.
2. From the root locus plot, obtain the x- and y-axis limits. Redefine the x-axis and y-axis limits. For example, refer to the following code

```
dt = 0.01;
omega = (-3:dt:3)*10^5;
sigma = (-1.5:dt:1.5)*10^6;
```

3. Declare the inputs to the meshgrid as follows:
```
[sigmagrid,omegagrid] = meshgrid(sigma,omega);
```

4. The complex frequency can be expressed as $s = \sigma + j\omega$. Therefore, define the complex frequency as

```
sgrid = sigmagrid + 1i*omegagrid;
```

5. Using the polyval function, declare the transfer function T_p and pass the inputs sigma and omega along with the numerator and denominator coefficients as follows:

```
Tp_3D = polyval(Tpnum,sgrid)./polyval(Tpden,sgrid);
```

6. Finally, obtain the 3D plot of the magnitude of the transfer function T_p on z-axis, variable sigma on the x-axis, and omega on the y-axis using the mesh command. The code is
```
mesh(sigma,omega,20*log10(abs(Tp_3D)));
```

7. Label the axes clearly to clearly differentiate between the real and imaginary frequencies.
8. The magnitude of T_p at the location of right-half plane zero is usually negative and maybe invisible. Use the rotate option on the figure window to view the characteristics at the location of the right-half plane.
9. Repeat the activity for $D = 0.85$. Observe the differences between the figures obtained for $D = 0.46$ and $D = 0.85$.

Post-lab Questions

1. What are the differences observed between the plots of the control-to-output transfer function for $D = 0.46$ and $D = 0.85$?
2. Describe the nature of the poles and zeros obtained in the plots. Based on the plots, how do the poles and zeros affect the characteristics of the transfer function. (You may explain with regard to the peaks and valleys, which occur at the pole-zero locations).
3. What are the consequences due to the movement of the right-half plane zero into the left-half plane beyond $D = 0.85$?

Note

This activity may be repeated as an additional activity for any transfer function discussed in the subsequent labs.

33

Open-Loop Input-to-Output Voltage Transfer Function of the Boost Converter in CCM

Objectives

The objectives of this lab are:

- To obtain theoretically the magnitude and phase plots of the input-to-output voltage transfer function M_v.
- To plot the response of the converter for a step change in the input voltage.
- To simulate the small-signal model on a circuit simulator and compare the theoretical and simulation results.

PART A: BODE PLOTS AND STEP RESPONSE USING MATLAB®

Theory

The small-signal input-to-output voltage transfer function M_v of the boost dc–dc converter in CCM in terms of the impedances is given by

$$M_v = \frac{v_o}{v_i} = \frac{1}{1 - D} \frac{Z_2}{Z_2 + \frac{Z_1}{(1-D^2)}}, \tag{33.1}$$

where

$$Z_1 = r + sL,$$

$$Z_2 = \frac{R_L \left(r_C + \frac{1}{sC} \right)}{R_L + r_C + \frac{1}{sC}}. \tag{33.2}$$

Laboratory Manual for Pulse-Width Modulated DC–DC Power Converters, First Edition.
Marian K. Kazimierczuk and Agasthya Ayachit.
© 2016 John Wiley & Sons, Ltd. Published 2016 by John Wiley & Sons, Ltd.

Substituting Equation (33.2) into (33.1), we get

$$M_v = \frac{(1-D)R_L r_C}{L(R_L + r_C)} \cdot \frac{\left(s + \frac{1}{Cr_C}\right)}{s^2 + s\left\{\frac{C[r(R_L+r_C)+(1-D)^2 R_L r_C]+L}{LC(R_L+r_C)}\right\} + \frac{r+(1-D)^2 R_L}{LC(R_L+r_C)}}. \tag{33.3}$$

In Equation (33.3), L is the inductance, C is the capacitance, R_L is the load resistance, D is the duty cycle, r_L is the equivalent series resistance (ESR) of the inductor, r_C is the ESR of the capacitor, and r is the equivalent averaged resistance given by

$$r = Dr_{DS} + (1-D)R_F + r_L, \tag{33.4}$$

where r_{DS} is the on-state resistance of the MOSFET and R_F is the on-resistance of the diode.

Specifications

Follow the specifications mentioned in Lab 31.

Quick Design

Assume an efficiency of 90%. Choose:

$L = 156$ μH, $r_L = 0.19$ Ω, $R_L = 64$ Ω, $D = D_{nom} = 0.46$, $C = 6.8$ μF, $r_C = 0.111$ Ω, $r_{DS} = 0.11$ Ω, $R_F = 15$ mΩ.

Procedure

A. Magnitude and Phase Plot of the Input-to-Output Voltage Transfer Function

1. Define all the parameters and variables on the MATLAB® editor window.
2. Define the $'s'$ function as s = tf('s'). This expression creates the variable $s = j\omega$ needed to define the transfer function in the Laplace domain.
3. Write the equation of the input-to-output voltage transfer function M_v in the impedance form given in Equation (33.1) or in the expanded form as given in Equation (33.3).
4. Define the command bode(Mv), in order to plot the magnitude and phase of the input-to-output voltage transfer function.
5. On the figure, right-click to change the x-axis from ω-domain in rad/s to frequency domain in Hz.

B. Step Response of the Input-to-Output Voltage Transfer Function

Write the following snippet of the code to plot the step response of the control-to-output voltage transfer function T_p.

```
t = 0:1e-7:3e-3;
u = step(Mv,t);
plot(t*1000,VO+u)
grid on
xlabel('{\it t} (ms)')
ylabel('{\it v_o} (V)')
```

In the above code, the time axis is multiplied by 1000 in order to represent the x-axis in milliseconds. Record the values of the maximum overshoot, undershoot, rise time, settling time, and steady-state value after step change.

PART B: SIMULATION OF THE CIRCUIT MODEL

C. Simulation of Small-Signal Model of the Boost Converter

1. The small-signal model of the boost converter needed to determine the input-to-output voltage transfer function is as shown in Figure 33.1. Construct the model on the circuit simulator. Make sure you have replaced the MOSFET with a current-controlled current source representing Di_l. Similarly, the diode must be replaced with a voltage-controlled voltage sources representing Dv_o.
2. Enter the values of the constants for both the dependent sources. To simulate the small-signal model at nominal operating conditions, let $D = D_{nom} = 0.46$, $R_L = R_{Lnom} = 64\ \Omega$.
3. Set the ac excitation due to the duty cycle d to zero. Provide an ac excitation for the small-signal perturbation at the input terminals, that is, the small-signal ac excitation is only due to the input voltage.
4. Set the simulation type to `ac analysis`. Let the `start frequency` be 0, `end/stop frequency` be 1 MHz. Choose an appropriate step size depending on the resolution needed. (E.g., 1000 points per decade). Perform the simulation.
5. Plot the magnitude and phase of the output voltage v_o. Compare the theoretical plots in Section B, item 3 with those obtained through simulations.

Post-lab Questions

1. What is the significance of the input-to-output voltage transfer function?
2. Prepare a table listing the values of (a) poles, (b) zeros, and (c) the dc and low-frequency gain obtained using the transfer function presented in Equation (33.3).

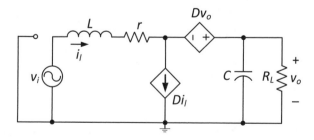

Figure 33.1 Small-signal model of the boost converter for determining the input-to-output voltage transfer function M_v with small-signal perturbations $d = 0$ and $i_o = 0$.

3. Draw an asymptotic Bode plot for the input-to-output voltage transfer function using the zero-pole-gain data obtained in the above question.
4. Determine the following values from the magnitude plot obtained in Section A:
   - corner frequency f_0
   - 3-dB bandwidth
   - unity-gain crossover frequency
   - dc and low-frequency gain
   - peak value of the gain
5. What are the differences between the input-to-output voltage transfer function and the control-to-output voltage transfer function? Express your views on the stability of the converter in open loop and the characteristics of the step response plot.

Note

The units of the x-axis of the Bode plot are preset as radians per second. The units may be changed to Hertz using the following command:

```
set(cstprefs.tbxprefs,'FrequencyUnits','Hz');
```

34

Open-Loop Small-Signal Input and Output Impedances of the Boost Converter in CCM

Objectives

The objectives of this lab are:

- To obtain theoretically the magnitude and phase plots of the input impedance Z_i and the output impedance Z_o.
- To simulate the small-signal model on the circuit simulator to determine the input and output impedances of the boost converter in CCM.

PART A: INPUT IMPEDANCE

Theory

The small-signal input impedance Z_i of the boost dc–dc converter in CCM in terms of the impedances is given as

$$Z_i = \frac{v_i}{i_i} = Z_1 + (1-D)^2 Z_2, \tag{34.1}$$

where

$$Z_1 = r + sL,$$

$$Z_2 = \frac{R_L \left(r_C + \frac{1}{sC} \right)}{R_L + r_C + \frac{1}{sC}}. \tag{34.2}$$

Laboratory Manual for Pulse-Width Modulated DC–DC Power Converters, First Edition.
Marian K. Kazimierczuk and Agasthya Ayachit.
© 2016 John Wiley & Sons, Ltd. Published 2016 by John Wiley & Sons, Ltd.

Substituting Equation (34.2) into (34.1), we get

$$Z_i = \frac{L\left\{s^2 + s\frac{C[r(R_L+r_C)+(1-D)^2R_Lr_C]+L}{LC(R_L+r_C)} + \frac{(1-D)^2R_L+r}{LC(R_L+r_C)}\right\}}{s + \frac{1}{C(R_L+r_C)}}. \tag{34.3}$$

In Equation (34.3), L is the inductance, C is the capacitance, R_L is the load resistance, D is the duty cycle, r_L is the equivalent series resistance (ESR) of the inductor, r_C is the ESR of the capacitor, and r is the equivalent averaged resistance given by

$$r = Dr_{DS} + (1-D)R_F + r_L, \tag{34.4}$$

where r_{DS} is the on-state resistance of the MOSFET and R_F is the on-resistance of the diode.

The small-signal output impedance of the boost converter in the impedance form is expressed as

$$Z_o = \frac{Z_1 Z_2}{Z_1 + Z_2(1-D)^2}. \tag{34.5}$$

Substituting Equation (34.2) into (34.5), we obtain

$$Z_o = \frac{R_L r_C}{R_L + r_C} \frac{\left(s + \frac{1}{Cr_C}\right)\left(s + \frac{r}{L}\right)}{s^2 + s\frac{C[r(R_L+r_C)+(1-D)^2R_Lr_C]+L}{LC(R_L+r_C)} + \frac{r+(1-D)^2R_L}{LC(R_L+r_C)}}. \tag{34.6}$$

Specifications

Follow the specifications mentioned in Lab 31.

Quick Design

Assume an efficiency of 90%. Choose:

$L = 156$ μH, $r_L = 0.19$ Ω, $R_L = 64$ Ω, $D = D_{nom} = 0.46$, $C = 6.8$ μF, $r_C = 0.111$ Ω, $r_{DS} = 0.11$ Ω, $R_F = 15$ mΩ.

Procedure

A. Magnitude and Phase Plot of the Input Impedance of the Boost Converter

1. Define all the parameters and variables on the MATLAB® editor window.
2. Define the $'s'$ function as s = tf('s'). This expression creates the variable $s = j\omega$ needed to define the transfer function in the Laplace domain.
3. Write the equation of the input impedance Z_i in the impedance form given in Equation (34.1) or in the expanded form as given in Equation (34.3).

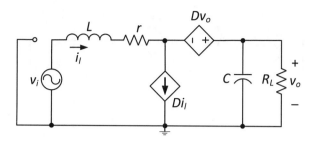

Figure 34.1 Small-signal model of the boost converter for determining the input impedance Z_i with small-signal perturbations $d = 0$ and $i_o = 0$.

4. Define the command `bode(Zi)` in order to plot the magnitude and phase of the input impedance.
5. On the figure, right-click to change the x-axis from ω-domain in `rad/s` to frequency domain in `Hz`.

B. Simulation of Small-Signal Model of the Boost Converter to Determine the Input Impedance

1. The small-signal model of the boost converter needed to determine the input impedance is as shown in Figure 34.1. Construct the model on the circuit simulator. The model is the same as that needed to determine the input-to-output voltage transfer function as discussed in Lab 35. Make sure you have replaced the MOSFET with a current-controlled current source representing Di_l. Similarly, the diode must be replaced with a voltage-controlled voltage source representing Dv_o.
2. Enter the values of the constants for both the dependent sources. To simulate the small-signal model at nominal operating conditions, let $D = D_{nom} = 0.46$, $R_L = R_{Lnom} = 64\ \Omega$.
3. Set the ac excitation due to the duty cycle d to zero. Provide an ac excitation for the small-signal perturbation at the input terminals, that is, the small-signal ac excitation is only due to the input voltage.
4. Set the simulation type to `ac analysis`. Let the `start frequency` be 0, `end/stop frequency` be 1 MHz. Choose an appropriate step size depending on the resolution needed. (E.g., 1000 points per decade). Perform the simulation.
5. Plot the magnitude and phase of the ratio of the input voltage v_i to the input current i_i. Compare the theoretical plots in Section B, item 3 with those obtained through simulations.

PART B: OUTPUT IMPEDANCE

C. Magnitude and Phase Plot of the Output Impedance of the Boost Converter

1. The expression for the output impedance is given in Equation (34.6).
2. Repeat the activities presented in Section A to obtain the theoretical plots for the output impedance of the boost converter.

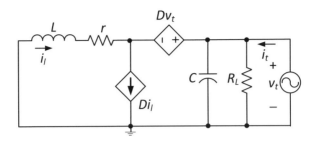

Figure 34.2 Small-signal model of the boost converter for determining the output impedance Z_o with small-signal perturbations $v_i = 0$ and $d = 0$.

D. Simulation of Small-Signal Model of the Boost Converter to Determine the Output Impedance

1. The small-signal model of the boost converter needed to determine the input impedance is as shown in Figure 34.2.
2. Perform the activities presented in Section B to determine the small-signal output impedance of the boost converter through simulations.

E. Open-Loop Response of the Output Voltage to Step Change in Load Current

The expression for the small-signal output impedance can be used to determine the open-loop response of the output voltage for a step change in the load current. Execute the following snippet of the code to obtain the step response.

```
t = 0:1e-7:3e-3;
u = step(Zo,t);
plot(t*1000,VO+u)
grid on
xlabel('{\it t} (ms)')
ylabel('{\it v_o} (V)')
```

In the code above, the time axis is multiplied by 1000 to represent the x-axis in milliseconds.

Post-lab Question

Explain the characteristics of the input and output impedances at dc and at high frequencies.

Note

The units of the x-axis of the Bode plot is preset as radians per second. The units may be changed to Hertz using the following command:

```
set(cstprefs.tbxprefs,'FrequencyUnits','Hz');
```

35

Feedforward Control of the Boost DC–DC Converter in CCM

Objective

The objective of this lab is to determine the dynamic performance of the boost dc–dc converter with input voltage feedforward control.

Theory

The dc voltage transfer function of the boost converter in continuous-conduction mode (CCM) is given by

$$\frac{V_O}{V_I} = \frac{\eta}{1-D} \Rightarrow D = 1 - \frac{\eta V_I}{V_O}, \tag{35.1}$$

where η is the overall efficiency of the boost converter, V_I is the input voltage, V_O is the output voltage, D is the duty cycle.

The dc component of the duty cycle is given by

$$D = 1 - \frac{V_R}{V_{Tm}}, \tag{35.2}$$

where V_R is the reference voltage to the inverting terminal of the comparator and V_{Tm} is the amplitude of the sawtooth voltage waveform. Rearranging Equation (35.2) and substituting Equation (35.1) into (35.2), we obtain

$$V_R = \frac{\eta V_{Tm} V_I}{V_O}. \tag{35.3}$$

The reference voltage V_R can be expressed as a fraction of the input voltage and is given by

$$V_R = \frac{R_B}{R_A + R_B} V_I. \tag{35.4}$$

Laboratory Manual for Pulse-Width Modulated DC–DC Power Converters, First Edition.
Marian K. Kazimierczuk and Agasthya Ayachit.
© 2016 John Wiley & Sons, Ltd. Published 2016 by John Wiley & Sons, Ltd.

Table 35.1 Parameters of the boost converter

Parameters	Notation	Value
Nominal dc input voltage	V_{Inom}	12 V
DC output voltage	V_O	20 V
Switching frequency	f_s	100 kHz
Maximum output current	I_{Omax}	0.5 A
Nominal output current	I_{Onom}	0.3125 A
Minimum output current	I_{Omin}	0.125 A
Output voltage ripple	V_r	$< 0.01 V_O$

Equating Equations (35.3) and (35.4), we get

$$\frac{R_B}{R_A + R_B} = \eta \frac{V_{Tm}}{V_O}. \tag{35.5}$$

Using Equation (35.5), the value of the resistances R_A and R_B can be determined.

Specifications

The specifications of the boost converter are as provided in Table 35.1.

Pre-lab

For the specifications provided in Tables 35.1 and 35.2, find the values of all the components and specifications of the dc–dc boost converter operating in CCM using the relevant design equations provided in Table A.1 in Appendix A.

Quick Design

Assume an efficiency of 90%. Choose:

$L = 156$ μH, ESR of the inductor $r_L = 0.19$ Ω, $R_{Lmin} = 40$ Ω, $R_{Lmax} = 160$ Ω, $D_{nom} = 0.46$, $C = 6.8$ μF, ESR of the capacitor $r_C = 0.111$ Ω. Using (35.3), the value of the reference voltage is $V_R = 4.86$ V. Let $R_A = 120$ kΩ, then using Equation (35.4), $R_B = 100$ kΩ.

MOSFET: International Rectifier IRF142 n-channel power MOSFET with $V_{DSS} = 100$ V, $I_{SM} = 24$ A, $r_{DS} = 0.11$ Ω at $T = 25°$C, $C_o = 100$ pF, and $V_t = 4$ V.

Diode: ON Semiconductor MBR10100 with $V_{RRM} = 100$ V, $I_F = 10$ A, $R_F = 15$ mΩ, and $V_F = 0.8$ V.

Operational amplifier: LM741 or any ideal op-amp with provision for power supply terminals.

Table 35.2 Parameters of the pulse-width modulator

Parameters	Notation	Value
Power supply voltage	$V_{CC}, -V_{EE}$	12 V, −12 V
Maximum amplitude of the triangular wave	V_{Tm}	9 V
Frequency of carrier wave	f_s	100 kHz

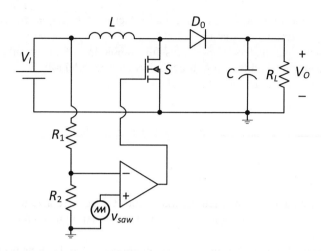

Figure 35.1 Circuit diagram of the PWM boost dc–dc converter with input voltage feedforward control.

Procedure

A. Simulation of the Boost Converter with Input Voltage Feedforward Control

1. Construct the circuit shown in Figure 35.1 on the circuit simulator.
2. Provide a sawtooth-wave voltage at the noninverting terminal of the op-amp. Provide the following details in order to generate the sawtooth voltage waveform: `initial voltage = 0 V`, `final/maximum voltage = 9 V`, `period = 10 μs`, `rise time = 9.9 μs`, `fall time = 0.1 μs`, and `ON time/width/duty cycle = 0.1 ns`.
3. Enter the values of all the other components. Make sure the selection of MOSFET is appropriate and meets the desired ratings.
4. Set simulation type to `transient analysis`. Set `end time = 20 ms` and a `time step = 0.1 μs`. Run the simulation.
5. Plot the following parameters after successful completion of the simulation. You may display the waveforms on different figure windows for better clarity.
   - Gate-to-source voltage v_{GS}, sawtooth voltage v_{saw}, and reference voltage V_R.
   - Output voltage v_O, output current i_O, and output power p_O.
   - Inductor current i_L.
6. Observe the inductor current waveform to ensure whether the current is in CCM. If the current is not in CCM, then increase the value of the inductor and repeat the simulation.

B. Simulation of the Boost Converter with Input Voltage Feedforward Control with Step Change in Input Voltage

1. Replace the input dc voltage source with a piecewise-linear voltage source. Define the following voltage profile:

Time (ms)	0	5.9	6	13.9	14	16.9
Input voltage V_I (V)	12	12	14	14	10	10

2. Set simulation type to `transient analysis`. Set `end time = 20 ms` and a `time step = 0.1 μs`. Run the simulation.
3. Plot the following parameters after successful completion of the simulation. You may display the waveforms on different figure windows for better clarity.
   - Output voltage v_O, output current i_O, and output power p_O.
   - Inductor current i_L.

Observe the transients which are generated at the instants of step change in the input voltage.

Post-lab Questions

1. What are the advantages of using feedforward control in dc–dc converters?
2. Explain the principle of operation of the boost converter with feedforward control in a few sentences.
3. What are the basic differences between feedforward control and feedback control?

36

P, PI, and PID Controller Design

Objectives

The objectives of this lab are:

- To design the basic proportional (P), proportional-integral (PI), and proportional-integral-derivative (PID) controllers.
- To observe the dynamic performance of these controllers for a square-wave input voltage.
- To realize these controllers on a simulation tool and to observe their time-domain characteristics.

Theory

Selection of the type of control scheme needed can be made based on the characteristics of the controller presented in Table 36.1.

Proportional Control: The implementation of the proportional (P) control using an operational amplifier is as shown in Figure 36.1. Assuming that the amplifier is configured to operate in the inverting mode, the input-to-output voltage gain of the controller is given by

$$K_p = \frac{v_{op}}{v_i} = -\frac{R_f}{R_s}. \tag{36.1}$$

Proportional-Integral Control: The circuit implementation of the proportional-integral (PI) control using an operational amplifier is as shown in Figure 36.2. Assuming that the amplifier is configured to operate in the inverting mode, the output voltage of the controller is given by

$$v_{oi} = -\left[\frac{R_f}{R_s}v_i + \frac{1}{R_iC_i}\int_0^t v_i dt\right] = -\left(K_p v_i + K_i \int_0^t v_i dt\right). \tag{36.2}$$

Laboratory Manual for Pulse-Width Modulated DC–DC Power Converters, First Edition.
Marian K. Kazimierczuk and Agasthya Ayachit.
© 2016 John Wiley & Sons, Ltd. Published 2016 by John Wiley & Sons, Ltd.

Table 36.1　Properties of the proportional, proportional integral and proportional derivative controllers

Type	Rise time	Overshoot	Settling time	Steady-state error
Proportional (K_p)	Decreases	Increases	No change	Decreases
Proportional-integral (K_i)	Decreases	Increases	Increases	Eliminated
Proportional-derivative (K_d)	No change	Decreases	Decreases	No change

Proportional-Derivative Control: The circuit implementation of the proportional-derivative (PD) control using an operational amplifier is as shown in Figure 36.3. Assuming that the amplifier is configured to operate in the inverting mode, the output voltage of the controller is given by

$$v_{od} = -\left[\frac{R_f}{R_s}v_i + R_dC_d\frac{dv_i}{dt}\right] = -\left(K_pv_i + K_d\frac{dv_i}{dt}\right). \tag{36.3}$$

The total output voltage of the PID controller can be determined by summing the output voltages due to the proportional, integral, and derivative stages. The total output voltage is

$$v_o = v_{op} + v_{od} + v_{oi} = -\left(\frac{R_f}{R_s} + \frac{1}{R_iC_i}\int_0^t v_i dt + R_dC_d\frac{dv_i}{dt}\right). \tag{36.4}$$

Specifications

Design a proportional controller having a gain $K_p = 2$, a proportional-integral controller with gain $K_i = 0.01$, and a proportional-derivative controller with gain $K_d = 100 \times 10^{-6}$.

Quick Design

For the proportional controller, let $R_f = 10$ kΩ and $R_s = 20$ kΩ.

For the integral control stage, let $R_i = 100$ MΩ and $C_i = 1$ μF.

For the derivative control stage, let $R_d = 1$ kΩ and $C_d = 0.1$ μF.

Let the gain of the summing amplifier be equal to 1 such that $R_1 = R_2 = R_3 = R_{fs} = 10$ kΩ.

NOTE: The circuit of the proportional controller is as shown in Figure 36.1. The operational features of the proportional controller is simple and must be obvious to the students.

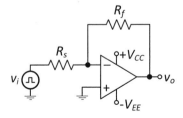

Figure 36.1　Op-amp implementation of the proportional controller.

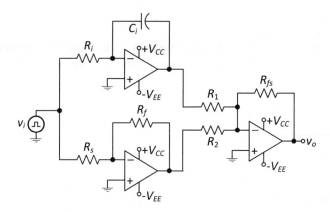

Figure 36.2   Op-amp implementation of the proportional-integral controller.

Procedure

A. Transient Analysis of the Proportional-Integral Controller

1. Construct the circuit of the PI controller as shown in Figure 36.2. Enter the values of all the components obtained using the given specifications.
2. Let the positive supply voltage to the op-amp be $V_{CC} = 12$ V. The negative power supply terminal $-V_{EE}$ can be connected to ground or to a voltage source of value -12 V.
3. Connect a pulse voltage source at the inputs to the two op-amps as shown in Figure 36.2. Set `time period = 10 ms`, `duty cycle/width = 0.5`, `initial voltage = 0` `V`, `amplitude = 0.5 V`. Let the `rise time` and `fall time` be equal to zero (optional).
4. Perform transient analysis. Set an `end time = 40 ms`. Provide a `time step = 0.01 ms`.

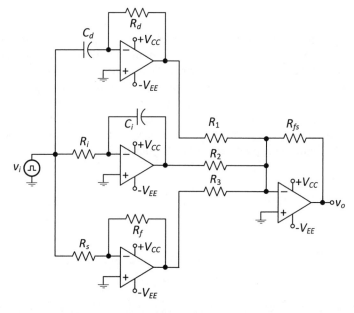

Figure 36.3   Op-amp implementation of the proportional-integral-derivative controller.

5. Plot the following:
   - Input voltage waveform.
   - Output voltage waveforms of the proportional controller and the integral controller.
   - Output voltage waveform of the summing amplifier representing the overall output voltage.
6. Observe the transient characteristics of the output voltage waveforms. Also, note the shape of the output voltage of the integral controller for a square-wave input voltage.

B. Transient Analysis of the Proportional-Integral-Derivative Controller

1. Construct the circuit of the PID controller as shown in Figure 36.3. Enter the values of all the components obtained using the given specifications.
2. Let the positive supply voltage to the op-amp be $V_{CC} = 12$ V. The negative power supply terminal $-V_{EE}$ can be connected to ground or to a voltage source of value -12 V.
3. Connect a pulse voltage source at the inverting terminals of the two op-amps. Set time period = 10 ms, duty cycle/width = 0.5, initial voltage = 0 V, amplitude = 0.5 V. Let the rise time and fall time be equal to zero (optional).
4. Perform transient analysis. Set an end time of 40 ms. Provide a time step of 0.01 ms.
5. Plot the following:
   - Input voltage waveform.
   - Output voltage waveform of the proportional, integral, and derivative controller stages.
   - Output voltage waveform of the summing amplifier representing the overall output voltage.
6. Observe the ringing in the output voltage of the summing amplifier generated by the derivative control stage.

Post-lab Questions

1. Design a PID controller with values: $K_p = 0.35, K_i = 25, K_d = 0.0017$. Simulate the converter and plot the output voltage waveform for an input defined in Sections A and B. The individual gains of the summing amplifier may be assumed to be unity.
2. Consider a square-wave input voltage waveform with a minimum value of 0 V and a maximum value of 1 V. Describe the shape of the output voltage waveform of the (a) proportional controller, (b) proportional-integral controller, and (c) proportional-derivative controller.
3. The proportional-derivative controller is associated with the noise problem and has adverse effects on high-frequency applications. Provide your reasoning.

37

P, PI, and PID Controllers: Bode and Transient Analysis

Objectives

The objectives of this lab are:

- To observe the various s-domain characteristics of the proportional, proportional-integral, and proportional-integral-derivative controllers.
- To consider a practical open-loop transfer function of the plant and modify its properties using the controller transfer function.

(Note: Students are encouraged to complete Lab 36, which involves the design and circuit-level implementation of these control schemes.)

Theory

The output voltage of the proportional-integral controller in the s-domain is expressed as

$$v_o(s) = K_p v_i(s) + K_i \frac{v_i(s)}{s} = v_i(s) K_p \left(\frac{s + \frac{K_i}{K_p}}{s} \right). \tag{37.1}$$

The output voltage of the proportional-derivative controller in the s-domain is expressed as

$$v_o(s) = K_p v_i(s) + s K_d v_i(s) = v_i(s) K_d \left(s + \frac{K_p}{K_d} \right). \tag{37.2}$$

Laboratory Manual for Pulse-Width Modulated DC–DC Power Converters, First Edition.
Marian K. Kazimierczuk and Agasthya Ayachit.
© 2016 John Wiley & Sons, Ltd. Published 2016 by John Wiley & Sons, Ltd.

The output voltage of the proportional-integral-derivative controller in the s-domain is expressed as

$$v_o(s) = K_p v_i(s) + K_i \frac{v_i(s)}{s} + s K_d v_i(s) = v_i(s) \left(\frac{s^2 K_d + s K_p + K_i}{s} \right). \tag{37.3}$$

From Equation (37.3), the input-to-output voltage transfer function can be expressed as

$$G(s) = \frac{v_o(s)}{v_i(s)} = \frac{s^2 K_d + s K_p + K_i}{s}, \tag{37.4}$$

indicating the presence of a pole at origin ($s = 0$) and two zeros at $s = \frac{-K_p \pm \sqrt{K_p^2 - 4K_d K_i}}{2K_d}$.

Example of the Open-Loop Transfer Function

Consider the input-to-output voltage transfer function of the boost converter operating in continuous-conduction mode (CCM) as given in Lab 33. For the values of the components provided in the specifications, one obtains the following transfer function.

$$M_v(s) = 383.2 \frac{s + (1.32 \times 10^6)}{s^2 + 5532s + (280.2 \times 10^6)}. \tag{37.5}$$

Procedure

A. Bode Analysis of the Transfer Functions Considering PID Controller

1. Define the transfer function of the PID controller as given in Equation (37.4) on MATLAB®. Let $K_p = 5$, $K_i = 1$, and $K_d = 20 \times 10^{-6}$. The transfer function can be conveniently represented as G = tf([Kd Kp Ki], [1 0]).
2. Define the example, that is, the input-to-output voltage transfer function M_v as given in Equation (37.5). The transfer function can be conveniently represented as

   Mv = 383.2*tf([1 (1.32e6)], [1 5532 (280.2e6)]).

3. Define the loop-gain transfer function, which indicates the stability of the open-loop converter with the PID controller. The loop-gain transfer function T can be expressed on MATLAB® as T = Mv*G.
4. Further, use the MATLAB's® built-in command to obtain the closed-loop input-to-output voltage transfer function M_{vcl}. The code to introduce feedback in the loop is Mvcl = feedback(T, 1), where 1 represents a feedback gain of unity.
5. Using the bode option on MATLAB®, obtain the following (on separate figure windows):
   - Bode plot of the open-loop input-to-output voltage transfer function $M_v(s)$.
   - Bode plot of the controller transfer function $G(s)$.
   - Bode plot of the loop-gain transfer function $T(s)$.
   - Bode plot of the closed-loop transfer function $M_{vcl}(s)$.
   - Bode plots of the open-loop converter transfer function $M_v(s)$ and the closed-loop transfer function $M_{vcl}(s)$.

B. Response of the System for Step Change in the Input Voltage

1. Define a time variable t as t = 0:1e-6:3e-3. The code to obtain step response has been discussed in Lab 31.
2. Use the built-in MATLAB® function step to obtain the step response of the transfer functions. An example code to execute the step response of input-to-output voltage transfer M_v is mv_step = step(Mv, t).
3. Similarly, define the code to obtain the step response of the closed-loop input-to-output voltage transfer function M_{vcl}.
4. Plot the two responses as a function of time (on the same figure window) using the MATLAB's® plot function.

Post-lab Questions

In Section A

1. Find the values of the (a) dc and low-frequency gain and (b) 3-dB cut-off frequency from the Bode plots of the input-to-output voltage transfer function M_v.
2. Find the values of the (a) dc and low-frequency gain, (b) 3-dB cut-off frequency, and (c) crossover frequency from the Bode plots of the loop-gain transfer function T. Also, estimate the gain margin (GM) and phase margin (PM) of the loop-gain transfer function.
3. Find the values of the (a) dc and low-frequency gain, (b) 3-dB cut-off frequency, and (c) crossover frequency from the Bode plots of the closed-loop input-to-output voltage transfer function M_{vcl}. Also, estimate the gain margin (GM) and phase margin (PM) of the transfer function M_{vcl}.
4. Provide a few comments on the stability of the converter in the open-loop and closed-loop configurations based on the gain and phase margin values.
5. What conclusions can be made on the bandwidth of the transfer functions M_v and M_{vcl}?

In Section B

1. Determine the settling time or the time taken for the response to reach steady state.
2. Determine the maximum overshoot of the two step response plots.
3. Determine the steady-state error of the two step response plots.

Additional Activity: Bode and Step Response Analysis using only PI Control

Repeat Sections A and B by letting $K_d = 0$. Identify the differences in the Bode and step response plots of the open-loop and closed-loop input-to-output voltage transfer functions. Also, measure the settling time, maximum overshoot, and steady-state error of the plots obtained. Comment on the effect of the PI and PID controllers on the open-loop transfer functions.

38

Transfer Functions of the Pulse-Width Modulator, Boost Converter Power Stage, and Feedback Network

Objectives

The objectives of this lab are:

- To determine the transfer function of the loop gain of the boost converter assuming a unity-gain control transfer function.
- To determine the frequency-domain characteristics of the loop-gain transfer function.

Theory

Figure 38.1 shows the block diagram representation of the closed-loop small-signal model of the boost converter with unity-gain control. The ac control voltage-to-duty cycle transfer function T_m of the pulse-width modulator is

$$T_m = \frac{d}{v_c} = \frac{1}{V_{Tm}}. \tag{38.1}$$

The control-to-output voltage transfer function $T_p(s)$ in terms of impedances is

$$T_p = \frac{v_o}{d} = \frac{V_O}{1-D} \frac{1 - \dfrac{Z_1}{(1-D^2)R_L}}{1 + \dfrac{Z_1}{(1-D^2)Z_2}}, \tag{38.2}$$

Laboratory Manual for Pulse-Width Modulated DC–DC Power Converters, First Edition.
Marian K. Kazimierczuk and Agasthya Ayachit.
© 2016 John Wiley & Sons, Ltd. Published 2016 by John Wiley & Sons, Ltd.

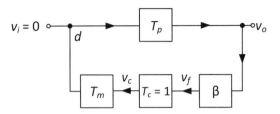

Figure 38.1 Block diagram representing the closed-loop small-signal model of the boost converter with unity-gain control.

where

$$Z_1 = r + sL,$$

$$Z_2 = \frac{R_L \left(r_C + \frac{1}{sC} \right)}{R_L + r_C + \frac{1}{sC}}. \tag{38.3}$$

Substituting Equation (38.3) into (38.2), we obtain the control-to-output voltage transfer function as

$$T_p = -\frac{V_O r_C}{(1-D)(R_L + r_C)} \frac{\left(s + \frac{1}{Cr_C} \right) \left\{ s - \frac{1}{L}[R_L(1-D)^2 - r] \right\}}{s^2 + s \left\{ \frac{C[r(R_L+r_C)+(1-D)^2 R_L r_C]+L}{LC(R_L+r_C)} \right\} + \frac{r+(1-D)^2 R_L}{LC(R_L+r_C)}}. \tag{38.4}$$

The transfer function of the feedback network β is

$$\beta = \frac{v_f}{v_o} = \frac{R_B}{R_A + R_B}. \tag{38.5}$$

Therefore, the loop-gain transfer function T_k of the pulse-width modulator, boost converter power stage, and feedback network is

$$T_k = \frac{v_f}{v_c} = \frac{v_f}{v_o} \times \frac{v_o}{d} \times \frac{d}{v_c} = \beta T_p T_m. \tag{38.6}$$

Substituting Equations (38.1), (38.4), and (38.5) into (38.6), the loop gain can be expressed as

$$T_k = \frac{\beta V_O r_C}{V_{Tm}(1-D)(R_L + r_C)} \frac{\left(s + \frac{1}{Cr_C} \right) \left\{ s - \frac{1}{L}[R_L(1-D)^2 - r] \right\}}{s^2 + s \left\{ \frac{C[r(R_L+r_C)+(1-D)^2 R_L r_C]+L}{LC(R_L+r_C)} \right\} + \frac{r+(1-D)^2 R_L}{LC(R_L+r_C)}}. \tag{38.7}$$

Specifications

Obtain the converter parameters and component values for the specifications of the boost converter provided in Lab 31.

Quick Design

Assume an efficiency of 90%. Choose:

$L = 156$ µH, $r_L = 0.19$ Ω, $R_L = 64$ Ω, $D = D_{nom} = 0.46$, $C = 6.8$ µF, $r_C = 0.111$ Ω, $r_{DS} = 0.11$ Ω, $R_F = 15$ mΩ.

Assume the maximum amplitude of the sawtooth waveform as $V_{Tm} = 4$ V. At $D = D_{nom}$, we have $V_F = V_C = D_{nom}V_{Tm}$. Therefore, $V_F = 1.84$ V. Also, $\beta = V_F/V_O = R_B/(R_A + R_B) = 1.84/20 = 0.0902$. Choose $R_A = 15$k Ω and $R_B = 1.5$ kΩ.

Procedure

A. Magnitude and Phase Plot of the Loop-Gain Transfer Function With Unity-Gain Control

1. Define all the parameters and variables on the MATLAB® editor window.
2. Define the $'s'$ function as $s = tf('s')$. This expression creates the variable $s = j\omega$ needed to define the transfer function in the Laplace domain.
3. Write the expressions of the transfer functions of the pulse-width modulator, control-to-output transfer function of the boost power stage, and the feedback network. The expressions are provided in Equations (38.1), (38.4), and (38.5). Define the loop-gain transfer function as given in Equation (38.6).
4. Define the command `bode(Tk)` in order to plot the magnitude and phase of the loop-gain transfer function.
5. On the figure, right-click to change the x-axis from ω-domain in `rad/s` to frequency domain in `Hz`.

B. Determination of Characteristics of the Loop-Gain Transfer Function

From the Bode plot of the loop-gain transfer function T_k obtained in Section A, determine the following parameters:

- The dc and low-frequency gain T_{k0} in decibels.
- The 3-dB cut-off frequency f_{3-dB} in Hertz.
- The crossover frequency f_c in Hertz.
- The gain margin GM in decibels and phase margin PM in degrees.

Post-lab Questions

1. What is the purpose of determining the loop-gain transfer function of the power converter in closed loop?

2. What comments can be made on the stability of the boost converter without any controller?
3. How do the transfer functions of the pulse-width modulator and the feedback network affect the loop-gain transfer function in terms of the dc gain and frequencies?
4. Tabulate all the values obtained in Section B. What conclusions can be made about the values of gain and phase margins? What are the desired values of gain and phase margins of any loop-gain transfer function in a power electronic converter needed to ensure stability?

Note

1. *Changing the units of x-axis*: The units of the *x*-axis of the Bode plot is preset as radians per second. The units may be changed to Hertz using the following command:

```
set(cstprefs.tbxprefs,'FrequencyUnits','Hz');
```

2. *Method to draw the magnitude and phase plots by predefining a frequency variable.*
   - Define a variable f. Let f vary from 0 to 1 MHz. Define the term s such that $s = 2\pi f$.
   - Define the expression of the small-signal transfer function. Include a dot $(\cdot)$ before every multiplication, division, or power symbol to enable element-wise operation. Make sure you include a dot in the definition of the term s also.
   - You may use the `plot` command. Let the *x*-axis be the frequency variable f. The *y*-axis must be the magnitude or the phase of the transfer function. In order to convert the magnitude of a transfer function *TF* from a linear scale to decibels, use `20*log10(TF)`. Similarly, to plot the phase in degrees, use `angle(TF)*180/pi`.

39

Closed-Loop Control-to-Output Voltage Transfer Function with Unity-Gain Control

Objectives

The objectives of this lab are:

- To determine the *s*-domain characteristics of the closed-loop control-to-output voltage transfer function with unity-gain control transfer function.
- To plot the step responses and compare the time-domain characteristics of open-loop and closed-loop control-to-output voltage transfer functions with unity-gain controller.

Theory

The closed-loop control-to-output voltage transfer function T_{pcl} with unity-gain control is given by

$$T_{pcl} = \frac{T_m T_p}{1 + \beta T_m T_p} = \frac{T_m T_p}{1 + T_k}, \tag{39.1}$$

where T_m is the gain of the pulse-width modulator, T_p is the open-loop control-to-output voltage transfer function, β is the gain of the feedback network, and $T_k = \beta T_m T_p$ is the loop gain of the converter assuming the controller gain to be equal to unity. The expressions for these quantities have been provided in Equations (38.1), (38.4), and (38.5) in Lab 38.

Specifications

Obtain the converter parameters and component values for the specifications of the boost converter provided in Lab 31.

Laboratory Manual for Pulse-Width Modulated DC–DC Power Converters, First Edition.
Marian K. Kazimierczuk and Agasthya Ayachit.
© 2016 John Wiley & Sons, Ltd. Published 2016 by John Wiley & Sons, Ltd.

Quick Design

Assume an efficiency of 90%. Choose:

$L = 156$ μH, $r_L = 0.19$ Ω, $R_L = 64$ Ω, $D = D_{nom} = 0.46$, $C = 6.8$ μF, $r_C = 0.111$ Ω, $r_{DS} = 0.11$ Ω, $R_F = 15$ mΩ.

Assume the maximum amplitude of the sawtooth waveform as $V_{Tm} = 4$ V. At $D = D_{nom}$, we have $V_F = V_C = D_{nom}V_{Tm}$. Therefore, $V_F = 1.84$ V. Also, $\beta = V_F/V_O = R_B/(R_A + R_B) = 1.84/20 = 0.0902$. Choose $R_A = 15$ kΩ and $R_B = 1.5$ kΩ.

Procedure

A. Magnitude and Phase Plot of the Closed-Loop Control-to-Output Voltage Gain Transfer Function With Unity-Gain Controller

1. Define all the parameters and variables on the MATLAB® editor window.
2. Define the $'s'$ function as s = tf('s'). This expression creates the variable $s = j\omega$ needed to define the transfer function in the Laplace domain.
3. Write the expression of the transfer function of the closed-loop control-to-output voltage as given in Equation (39.1).
4. Define the command bode(Tpcl), in order to plot the magnitude and phase of the closed-loop control-to-output voltage transfer function. On the same figure window, also plot the open-loop control-to-output voltage transfer function as defined in Equation (38.2) in Lab 38.
5. On the figure, right-click to change the x-axis from ω-domain in rad/s to frequency domain in Hz.

B. Response for Step Change in the Duty Cycle

1. Define a time variable t as t = 0:1e-6:3e-3.
2. Use the built-in MATLAB® function step to obtain the step response of the transfer functions. For example, the code to execute the step response of control-to-output voltage transfer T_p is Tp_step = step(Tp, t). The complete code to plot the step response has been provided in Lab 31.
3. Plot the step response of the following on the same figure window using the MATLAB's plot function:
   - Open-loop control-to-output voltage transfer function T_p.
   - Closed-loop input-to-output voltage transfer function T_{pcl}.

Post-lab Questions

1. What does negative feedback do to the stability of the converter?
2. What differences do you observe between the closed-loop and open-loop control-to-output voltage transfer functions obtained in this Lab and Lab 31, respectively?
3. Measure the following values from the magnitude plots obtained in Section A:
   - Corner frequency f_0 in Hz
   - 3-dB bandwidth in Hz

- Unity-gain crossover frequency in Hz
- DC and low-frequency gain in dB
- Peak value of the gain in dB

Tabulate the values for better readability.

4. What are the characteristics of the poles and zeros in the open-loop and closed-loop control-to-output voltage transfer functions?

5. Explain the differences between the step response plots obtained in Section B. Tabulate the values of:
    - steady-state error
    - steady-state values of the output voltage and duty cycle
    - rise time
    - settling time

Additional Activity: Repeat Lab 32 to determine the characteristics of the T_{pcl} transfer function.

40

Simulation of the Closed-Loop Boost Converter with Proportional Control

Objective

The objective of this lab is to perform transient analysis on the closed-loop boost converter with proportional control using a simulation tool.

Specifications

Obtain the converter parameters and component values for the specifications of the boost converter provided in Lab 31.

Quick Design

Assume an efficiency of 90%. Choose:

$L = 156$ μH, $r_L = 0.19$ Ω, $R_L = 64$ Ω, $D = D_{nom} = 0.46$, $C = 6.8$ μF, $r_C = 0.111$ Ω, $r_{DS} = 0.11$ Ω, $R_F = 15$ mΩ.

Assume the maximum amplitude of the sawtooth waveform as $V_{Tm} = 4$ V. At $D = D_{nom}$, we have $V_F = V_C = D_{nom}V_{Tm}$. Therefore, $V_F = 1.84$ V. Also, $\beta = V_F/V_O = R_B/(R_A + R_B) = 1.84/20 = 0.0902$. Choose $R_A = 15$ kΩ and $R_B = 1.5$ kΩ.

MOSFET: International Rectifier IRF142 n-channel power MOSFET with $V_{DSS} = 100$ V, $I_{SM} = 24$ A, $r_{DS} = 0.11$ Ω at $T = 25°C$, $C_o = 100$ pF, and $V_t = 4$ V.

Diode: ON Semiconductor MBR10100 with $V_{RRM} = 100$ V, $I_F = 10$ A, $R_F = 15$ mΩ, and $V_F = 0.8$ V.

Laboratory Manual for Pulse-Width Modulated DC–DC Power Converters, First Edition.
Marian K. Kazimierczuk and Agasthya Ayachit.
© 2016 John Wiley & Sons, Ltd. Published 2016 by John Wiley & Sons, Ltd.

Operational amplifier: LM741 or any ideal op-amp with provision for power supply terminals.

Proportional controller (or error amplifier): Let the value of the feedback resistor be $R_F = R_B = 1.5$ kΩ.

Procedure

A. Transient Analysis of the Closed-Loop Boost Converter

1. Construct the circuit of the closed-loop boost dc–dc converter as shown in Figure 40.1 on the circuit simulator. Enter the values of all the components present in the feedback network, controller, pulse-width modulator, and the boost converter stages.
2. Provide a sawtooth-wave voltage at the inverting terminal of the modulator op-amp. Provide the following details in order to generate the sawtooth voltage waveform: `initial voltage = 0 V`, `final/maximum voltage = 4 V`, `period = 10 μs`, `rise time = 9.9 μs`, `fall time = 0.1 μs`, and `ON time/width/duty cycle = 0.1 ns`.
3. Connect the feedback voltage to the inverting terminal of the error amplifier. Connect a dc voltage source of value $V_R = V_F = 1.84$ V to the noninverting terminal of the error amplifier.
4. Connect the output terminal of the error amplifier to the noninverting terminal of the pulse-width modulator.
5. Set simulation type to `transient analysis`. Set `end time = 10 ms` and `time step = 0.1 μs`. Run the simulation.
6. Plot the following parameters after successful completion of the simulation. You may display the waveforms on different figure windows for better clarity.
   - Gate-to-source voltage v_{GS} and drain-to-source v_{DS}.
   - Error voltage v_e, sawtooth voltage v_{saw}, and control voltage V_C.
   - Output voltage v_O, output current i_O, and output power p_O.
   - Inductor current i_L.
7. Observe the inductor current waveform to ensure whether the current is in CCM. If the current is not in CCM, then increase the value of the inductor and repeat the simulation.

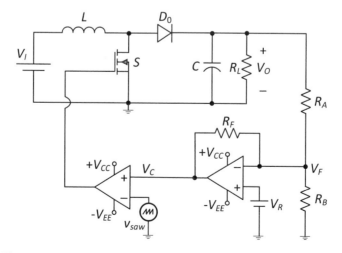

Figure 40.1   Circuit diagram of the closed-loop dc–dc boost converter.

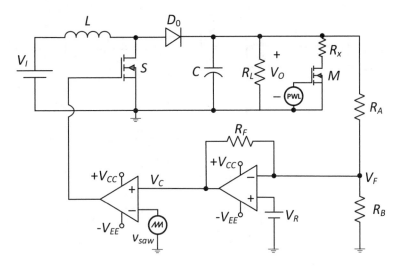

Figure 40.2 Circuit of the closed-loop boost dc–dc converter used to simulate step change in load resistance.

B. Simulation of the Closed-Loop Boost Converter for Step Change in the Load Resistance

1. Construct the circuit shown in Figure 40.2 on the circuit simulator. The circuit consists of a series connection of a resistor R_x and an ideal MOSFET M and is placed in parallel with the load resistor with the load resistor R_L. Let $R_x = 40\ \Omega$. This is needed in order to simulate the step change in the load resistance with respect to time.
2. Connect a piecewise-linear voltage source to the gate of the MOSFET M. Enter the details in Table 40.1 in order to provide a time-varying voltage to the gate of M_2.
3. Set simulation type to `transient analysis`. Set `end time` = `20 ms` and `time step` = `0.1 μs`. Run the simulation.
4. Plot the following parameters after successful completion of the simulation. You may display the waveforms on different figure windows for better clarity.
   - Output voltage v_O, output current i_O, and output power p_O.
   - Inductor current i_L.
   Observe the transient characteristics of the above waveforms. Verify whether the output voltage is maintained at $V_O \approx 20$ V for the entire duration of the simulation time.

C. Simulation of the Closed-Loop Boost Converter for Step Change in the Input Voltage

1. Construct the circuit shown in Figure 40.3 on the circuit simulator.

Table 40.1 Gate voltage profile of the MOSFET M

Time (ms)	0	5.99	6	13.9	14
Gate voltage V_G (V)	0	0	10	10	0

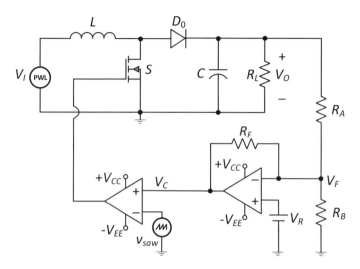

Figure 40.3 Circuit of the closed-loop boost dc–dc converter used to simulate step change in input voltage.

Table 40.2 Input voltage profile for the PWL voltage source

Time (ms)	0	5.9	6	13.9	14	16.9
Input voltage V_I (V)	12	12	14	14	10	10

2. Replace the constant dc supply voltage at the input of the boost converter with a piecewise-linear voltage source. Enter the details in Table 40.2 in order to obtain a time-varying input voltage waveform.
3. Let the parameters of the sawtooth-wave voltage source remain the same as those provided in Section A.
4. Set simulation type to `transient analysis`. Set `end time = 20 ms` and `time step = 0.1 μs`. Run the simulation.
5. Plot the following parameters after successful completion of the simulation. You may display the waveforms on different figure windows for better clarity.
   - Output voltage v_O, output current i_O, and output power p_O.
   - Inductor current i_L.
   Observe the transient characteristics of the above waveforms. Verify whether the output voltage is maintained at $V_O \approx 20$ V for the entire duration of the simulation time.

Post-lab Question

To what degree does the proportional controller for the boost converter in CCM satisfy the gain and phase margin requirements?

41

Voltage-Mode Control of Boost DC–DC Converter with Integral-Double-Lead Controller

Objective

The objective of this lab is to perform transient analysis on the voltage-mode controlled boost converter with the integral-double-lead control using a simulation tool.

Specifications

Obtain the parameters of the boost converter provided in Lab 31.

Quick Design

Assume an efficiency of 90%. Choose:

$L = 156$ μH, $r_L = 0.19$ Ω, $R_L = 64$ Ω, $D = D_{nom} = 0.46$, $C = 6.8$ μF, $r_C = 0.111$ Ω, $r_{DS} = 0.11$ Ω, $R_F = 15$ mΩ.

MOSFET: International Rectifier IRF142 n-channel power MOSFET with $V_{DSS} = 100$ V, $I_{SM} = 24$ A, $r_{DS} = 0.11$ Ω at $T = 25°$C, $C_o = 100$ pF, and $V_t = 4$ V.

Diode: ON Semiconductor MBR10100 with $V_{RRM} = 100$ V, $I_F = 10$ A, $R_F = 15$ mΩ, and $V_F = 0.8$ V.

Operational amplifier: Use LM741 or any ideal op-amp. Let the supply voltage for pulse-width modulator be $V_{CC1} = 12$ V and supply voltage for the controller be $V_{CC2} = 3.3$ V.

Assume the maximum amplitude of the sawtooth waveform as $V_{Tm} = 5$ V. Let the reference voltage $V_R = 2.5$ V. Therefore, $\beta = V_R/V_O = R_B/(R_A + R_B) = 2.5/20 = 0.125$. Choose $R_A = 4.3$ kΩ and $R_B = 620$ Ω.

Integral-double-lead controller: Let $R_1 = 100$ kΩ, $R_2 = 100$ kΩ, and $R_3 = 2.2$ kΩ. Pick $C_1 = 5$ nF, $C_2 = 150$ pF, and $C_3 = 5$ nF.

Laboratory Manual for Pulse-Width Modulated DC–DC Power Converters, First Edition.
Marian K. Kazimierczuk and Agasthya Ayachit.
© 2016 John Wiley & Sons, Ltd. Published 2016 by John Wiley & Sons, Ltd.

Procedure

A. *Transient Analysis of the Closed-Loop Boost Converter*

1. Construct the circuit of the closed-loop boost dc–dc converter as shown in Figure 41.1 on the circuit simulator. Enter the values of all the components present in the feedback, controller, pulse-width modulator, and the boost converter stages.
2. Provide a sawtooth-wave voltage at the inverting terminal of the op-amp. Provide the following details in order to generate the sawtooth voltage waveform: `initial voltage = 0 V`, `final/maximum voltage = 5 V`, `period = 10 µs`, `rise time = 9.9 µs`, `fall time = 0.1 µs`, and `ON time/width/duty cycle = 0.1 ns`.
3. Connect the feedback voltage to the inverting terminal of the error amplifier. Connect a dc voltage source of value $V_R = 2.5$ V to the noninverting terminal of the error amplifier.
4. Connect the output terminal of the error amplifier to the noninverting terminal of the pulse-width modulator.
5. Set simulation type to `transient analysis`. Set `end time = 10 ms` and `time step = 0.1 µs`. Run the simulation.
6. Plot the following parameters after successful completion of the simulation. You may display the waveforms on different figure windows for better clarity.
   - Gate-to-source voltage v_{GS} and drain-to-source voltage v_{DS}.
   - Error voltage v_e, sawtooth voltage v_{saw}, and control voltage V_C.
   - Output voltage v_O, output current i_O, and output power p_O.
   - Inductor current i_L.
7. Observe the inductor current waveform to ensure whether the current is in CCM. If the current is not in CCM, then increase the value of the inductor and repeat the simulation.

B. *Simulation of the Closed-Loop Boost Converter for Step Change in the Load Resistance*

1. Construct the circuit shown in Figure 41.2 on the circuit simulator. The circuit consists of a series connection of a resistor R_x and an ideal MOSFET M and is placed in parallel with

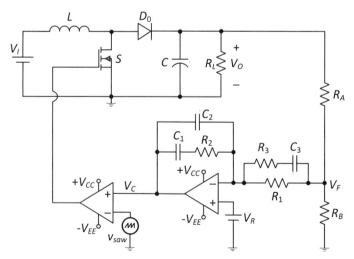

Figure 41.1   Circuit diagram of the voltage-mode controlled dc–dc boost converter.

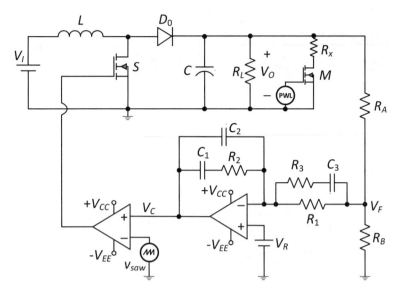

Figure 41.2   Circuit of the voltage-mode controlled boost dc–dc converter used to simulate step change in load resistance.

the load resistor R_L. Let $R_x = 40\ \Omega$. This is needed in order to simulate the step change in the load resistance with respect to time.

2. Provide a piecewise-linear voltage to the gate of the MOSFET M. Enter the details in Table 41.1 in order to provide a time-varying voltage to the gate of M.
3. Set simulation type to `transient analysis`. Set `end time = 20 ms` and `time step = 0.1 μs`. Run the simulation.
4. Plot the following parameters after successful completion of the simulation. You may display the waveforms on different figure windows for better clarity.
   - Gate-to-source voltage v_{GS} and drain-to-source voltage v_{DS}.
   - Error voltage v_e, sawtooth voltage v_{saw}, and control voltage V_C.
   - Output voltage v_O, output current i_O, and output power p_O.
   - Inductor current i_L.

   Observe the transient characteristics of the above waveforms. Verify whether the output voltage is maintained at $V_O \approx 20$ V for the entire duration of the simulation time.

C. Simulation of the Closed-Loop Boost Converter for Step Change in the Input Voltage

1. Construct the circuit shown in Figure 41.3 on the circuit simulator.
2. Provide a piecewise-linear voltage source in place of the input dc voltage to the boost converter. Enter the details in Table 41.2 in order to obtain a time-varying input voltage waveform.

Table 41.1   Gate voltage profile of the MOSFET M

Time (ms)	0	5.99	6	13.9	14
Gate voltage V_G (V)	0	0	10	10	0

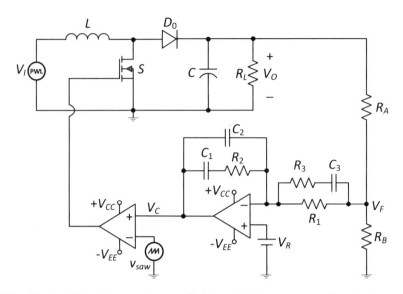

Figure 41.3 Circuit of the voltage-mode controlled dc–dc converter used to simulate step change in input voltage.

Table 41.2   Input voltage profile for the PWL voltage source

Time (ms)	0	5.9	6	13.9	14	16.9
Input voltage V_I (V)	12	12	14	14	10	10

3. Set simulation type to `transient analysis`. Set `end time = 20 ms` and `time step = 0.1 µs`. Run the simulation.
4. Plot the following parameters after successful completion of the simulation. You may display the waveforms on different figure windows for better clarity.
   - Gate-to-source voltage v_{GS} and drain-to-source voltage v_{DS}.
   - Error voltage v_e, sawtooth voltage v_{saw}, and control voltage V_C.
   - Output voltage v_O, output current i_O, and output power p_O.
   - Inductor current i_L.

   Observe the transient characteristics of the above waveforms. Verify whether the output voltage is maintained at $V_O \approx 20$ V for the entire duration of the simulation time.

Post-lab Question

Provide a detailed design methodology for the integral-double-lead controller shown in Fig. 41.1 for the boost converter in CCM.

42

Control-to-Output Voltage Transfer Function of the Open-Loop Buck DC–DC Converter

Objectives

The objectives of this lab are:

1. To analyze the open-loop control-to-output voltage and open-loop input-to-output voltage transfer functions of the buck converter in continuous-conduction mode (CCM).
2. To analyze the step response characteristics of the open-loop buck converter.

Theory

The small-signal control-to-output voltage transfer function T_p of the buck dc–dc converter in CCM is given by

$$T_p = \frac{v_o}{d} = \frac{V_I r_C R_L}{L(R_L + r_C)} \frac{s + \omega_Z}{s^2 + 2\xi\omega_0 s + \omega_0^2} \tag{42.1}$$

where

$$\omega_Z = \frac{1}{r_C C},$$

$$\omega_0 = \sqrt{\frac{r + R}{LC(R_L + r_C)}},$$

Laboratory Manual for Pulse-Width Modulated DC–DC Power Converters, First Edition.
Marian K. Kazimierczuk and Agasthya Ayachit.
© 2016 John Wiley & Sons, Ltd. Published 2016 by John Wiley & Sons, Ltd.

and

$$\xi = \frac{C(R_L r + r_C R_L + r_C r)}{2\sqrt{LC(R_L + r)(R_L + r_C)}}.$$
(42.2)

In Equation (42.1), V_I is the input voltage of the converter, L is the inductance, C is the capacitance, R_L is the load resistance, r_L is the equivalent series resistance (ESR) of the inductor, r_C is the ESR of the capacitor, and r is the equivalent averaged resistance given by

$$r = Dr_{DS} + (1 - D)R_F + r_L,$$
(42.3)

where r_{DS} is the on-state resistance of the MOSFET and R_F is the forward resistance of the diode.

Specifications

The specifications of the buck converter are as given in Table 42.1.

Pre-lab

For the specifications provided, find the values of all the components and specifications for the buck dc–dc converter operating in CCM using the relevant design equations provided Table A.1 in Appendix A.

Quick Design

Assume an overall efficiency of $\eta = 90\%$. Choose:
$L = 301$ μH, $R_{Lmin} = 10$ Ω, $D_{nom} = 0.555$, $C = 51.2$ μF, the equivalent averaged resistance $r = 0.16$ Ω.

Table 42.1 Parameters and their values

Parameters	Notation	Value
Nominal dc input voltage	V_I	28 V
DC output voltage	V_O	14 V
Switching frequency	f_s	100 kHz
Maximum output current	I_{Omax}	1.4 A
Minimum output current	I_{Omin}	0.5 A
Output voltage ripple	V_r	$< 0.01 V_O$

Procedure

A. Magnitude and Phase Plot of the Control-to-Output Voltage Transfer Function

1. Define all the parameters and variables on the MATLAB® editor window.
2. Define the '*s*' function as `s = tf('s')`. This expression creates the variable $s = j\omega$ needed to define the transfer function in the Laplace domain.
3. Write the equation of the control-to-output voltage transfer function T_p given in Equation (42.1).
4. Define the command `bode(Tp)`, in order to plot the magnitude and phase of the control-to-output voltage transfer function.
5. On the figure, right-click to change the *x*-axis from ω-domain in `rad/s` to frequency domain in `Hz`.

B. Response of the Buck Converter for Step Change in the Duty Cycle

Write the following snippet of the code to plot the step response of the control-to-output voltage transfer function T_p.

```
t = 0:1e-7:3e-3;
u = step(Tp,t);
plot(t*1000,VO + (0.1*u))
grid on
xlabel('{\it t} (ms)')
ylabel('{\it v_O} (V)')
```

In the above code, the time axis is multiplied by 1000 in order to represent the *x*-axis in milliseconds.

Post-lab Questions

1. What are the characteristics and significance of the control-to-output voltage transfer function of the buck converter?
2. What is the effect of the duty cycle on the location of poles and zeros of the control-to-output voltage transfer function?
3. Solve Equation (42.1) to determine the poles and zeros of the transfer function. Prepare a table listing the values of (a) poles, (b) zeros, and (c) the dc and low-frequency gain.
4. Draw an asymptotic Bode plot for the control-to-output voltage transfer function using the zero-pole-gain data obtained in the above question.
5. Determine the following values from the magnitude plot obtained in Section A:
   - Corner frequency f_0
   - 3-dB bandwidth
   - Unity-gain crossover frequency
   - DC and low-frequency gain
   - Peak value of the gain

6. Determine the expressions for the poles and zeros from the control-to-output voltage transfer function. Verify if the values of the poles and zeros obtained theoretically is in accordance with those obtained through MATLAB®.
7. In Section B, determine the values of the following quantities:
   - steady-state voltage value after step change
   - steady-state error
   - rise time
   - settling time
   - maximum overshoot
8. From the plot of step response, observe whether an undershoot is present in the output voltage waveform at $t = 0$. What are the differences between the step response plots of the buck and boost dc–dc converters?

Note

1. *Changing the unit of x-axis*: The unit of the x-axis of the Bode plot is preset as radians per second. The unit may be changed to Hertz using the following command:

   ```
   set(cstprefs.tbxprefs,'FrequencyUnits','Hz');
   ```

2. *Method to draw the magnitude and phase plots by predefining a frequency variable.*
   - Define a variable f. Let f vary from 0 to 1 MHz. Define the term s such that $s = 2\pi f$.
   - Define the expression of the small-signal transfer function. Include a dot ($\cdot$) before every multiplication, division, or power symbol to enable element-wise operation. Make sure you include a dot in the definition of the term s also.
   - You may use the `plot` command. Let the x-axis be the frequency variable f. The y-axis must be the magnitude or the phase of the transfer function. In order to convert the magnitude of a transfer function TF from a linear scale to decibels, use `20*log10(TF)`. Similarly, to plot the phase in degrees, use `angle(TF)*180/pi`.

43

Voltage-Mode Control of Buck DC–DC Converter

Objective

The objective of this lab is to perform transient analysis on the voltage-mode controlled buck converter with second-order proportional integral-single-lead control using a simulation tool.

Specifications

Obtain the parameters of the buck converter provided in Lab 42.

Quick Design

Assume an overall efficiency of $\eta = 90\%$. Choose:

$L = 301$ µH, $R_{Lmin} = 10$ Ω, $D_{nom} = 0.555$, $C = 51.2$ µF, the equivalent averaged resistance $r = 0.16$ Ω.

MOSFET: International Rectifier IRF142 n-channel power MOSFET with $V_{DSS} = 100$ V, $I_{SM} = 24$ A, $r_{DS} = 0.11$ Ω at $T = 25°C$, $C_o = 100$ pF, and $V_t = 4$ V.

Diode: ON Semiconductor MBR10100 with $V_{RRM} = 100$ V, $I_F = 10$ A, $R_F = 15$ mΩ, and $V_F = 0.8$ V.

Operational amplifier: Use LM741 or any ideal op-amp. Let the supply voltage for pulse-width modulator be $V_{CC1} = 12$ V and supply voltage for the controller be $V_{CC2} = 10$ V. The V_{EE} terminals of both the op-amps can be connected to ground.

Assume the maximum amplitude of the sawtooth waveform as $V_{Tm} = 10$ V. Let the reference voltage $V_R = 5$ V. Therefore, $\beta = V_R/V_O = R_B/(R_A + R_B) = 5/14 = 0.3571$. Choose $R_A = 1.2$ kΩ and $R_B = 620$ Ω.

Integral-single-lead controller: Let $R_1 = 1$ kΩ and $R_2 = 110$ kΩ. Pick $C_1 = 4.7$ nF and $C_2 = 6.8$ pF.

Laboratory Manual for Pulse-Width Modulated DC–DC Power Converters, First Edition.
Marian K. Kazimierczuk and Agasthya Ayachit.
© 2016 John Wiley & Sons, Ltd. Published 2016 by John Wiley & Sons, Ltd.

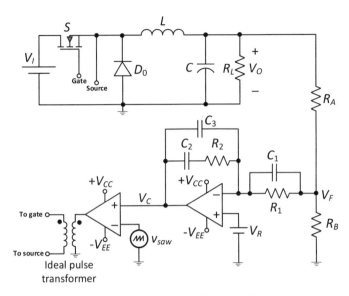

Figure 43.1 Circuit diagram of the voltage-mode controlled dc–dc buck converter with integral-single-lead control.

Procedure

A. Transient Analysis of the Voltage-Mode Controlled Buck Converter

1. Construct the circuit of the closed-loop buck dc–dc converter as shown in Figure 43.1 on the circuit simulator. Enter the values of all the components present in the feedback, controller, pulse-width modulator, and the buck converter stages.
2. You may use an ideal transformer as an interface between the single-ended output terminal of the op-amp and the gate-to-source terminal of the MOSFET. The Spice code of the ideal transformer as a subcircuit is given at the end of the handout. Alternatively, a dc/dc transformer with unity turns ratio can also be used.
3. Provide a sawtooth-wave voltage at the inverting terminal of the op-amp as pulse-width modulator. Provide the following details in order to generate the sawtooth voltage waveform: `initial voltage = 0 V, final/maximum voltage = 10 V, period = 10 μs, rise time = 9.9 μs, fall time = 0.1 μs`, and `ON time/width/duty cycle = 0.1 ns`.
4. Connect the feedback voltage to the inverting terminal of the error amplifier. Connect a dc voltage source of value $V_R = 5$ V to the noninverting terminal of the error amplifier.
5. Connect the output terminal of the error amplifier to the noninverting terminal of the pulse-width modulator.
6. Set simulation type to `transient analysis`. Set `end time = 10 ms` and `time step = 0.1 μs`. Run the simulation.
7. Plot the following parameters after successful completion of the simulation. You may display the waveforms on different figure windows for better clarity.
   - Gate-to-source voltage v_{GS} and drain-to-source voltage v_{DS}.
   - Error voltage v_e, sawtooth voltage v_{saw}, and control voltage V_C.
   - Output voltage v_O, output current i_O, and output power p_O.
   - Inductor current i_L.

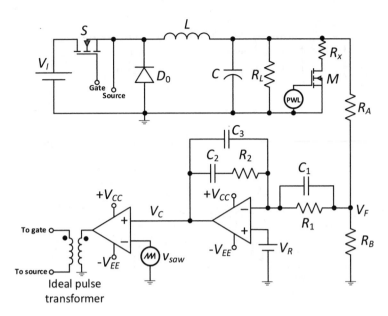

Figure 43.2 Circuit of the voltage-mode controlled buck dc–dc converter used to simulate step change in load resistance.

8. Observe the inductor current waveform to ensure whether the current is in CCM. If the current is not in CCM, then increase the value of the inductor and repeat the simulation.

B. Simulation of the Closed-Loop Buck Converter for Step Change in the Load Resistance

1. Construct the circuit shown in Figure 43.2 on the circuit simulator. The circuit consists of a series connection of a resistor R_x and an ideal MOSFET M and is placed in parallel with the load resistor R_L. Let $R_x = 10\ \Omega$. This is needed in order to simulate the step change in the load resistance with respect to time.
2. Connect a piecewise-linear voltage to the gate of the MOSFET M. Enter the details in Table 43.1 in order to provide a time-varying voltage to the gate of M.
3. Set simulation type to `transient analysis`. Set end time = 20 ms and time step = 0.1 μs. Run the simulation.
4. Plot the following parameters after successful completion of the simulation. You may display the waveforms on different figure windows for better clarity.
   - Gate-to-source voltage v_{GS} and drain-to-source voltage v_{DS}.
   - Error voltage v_e, sawtooth voltage v_{saw}, and control voltage V_C.
   - Output voltage v_O, output current i_O, and output power p_O.
   - Inductor current i_L.

Table 43.1 Gate voltage profile of the MOSFET M

Time (ms)	0	5.99	6	13.9	14
Gate voltage V_G (V)	0	0	10	10	0

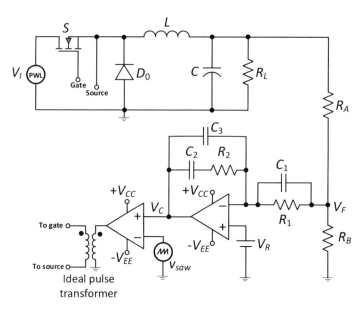

Figure 43.3   Circuit of the closed-loop boost dc–dc converter used to simulate step change in input voltage.

Observe the transient characteristics of the above waveforms. Verify whether the output voltage is maintained at $V_O \approx 14$ V for the entire duration of the simulation time.

C. Simulation of the Closed-Loop Buck Converter for Step Change in the Input Voltage

1. Construct circuit shown in Figure 43.3 on the circuit simulator.
2. Replace the constant dc voltage supply at the input to the buck converter with a piecewise-linear voltage source. Enter the details in Table 43.2 in order to provide a time-varying input voltage waveform.
3. Set simulation type to `transient analysis`. Set `end time = 20 ms` and `time step = 0.1 µs`. Run the simulation.
4. Plot the following parameters after successful completion of the simulation. You may display the waveforms on different figure windows for better clarity.
   - Gate-to-source voltage v_{GS} and drain-to-source voltage v_{DS}.
   - Error voltage v_e, sawtooth voltage v_{saw}, and control voltage V_C.
   - Output voltage v_O, output current i_O, and output power p_O.
   - Inductor current i_L.

   Observe the transient characteristics of the above waveforms. Verify whether the output voltage is maintained at $V_O \approx 14$ V for the entire duration of the simulation time.

Table 43.2   Input voltage profile for the PWL voltage source

Time (ms)	0	5.9	6	13.9	14	16.9
Input voltage V_I **(V)**	28	28	30	30	26	26

Post-lab Question

Explain the operation of the buck converter with the proportional controller in the feedback loop.

Note

The Spice code of ideal transformer is as follows:

```
.subckt UE_Ideal_Xformer P+ P- S+ S-
E1 S+ S- P+ P- {NS/NP}
F1 P- P+ E1 {NS/NP}
.param NP=1 NS=1
.backanno
.ends
```

44

Feedforward Control of the Buck DC–DC Converter in CCM

Objective

The objective of this lab is to determine the dynamic performance of the buck dc–dc converter with input voltage feedforward control.

Specifications

The specifications of the buck converter are as given in Table 44.1.

Pre-lab

For the specifications provided, find the values of all the components and parameters of the buck dc–dc converter operating in CCM using the design equations provided in Table A.1 in Appendix A.

Quick Design

Assume an overall efficiency of $\eta = 90\%$. Choose:

$L = 301$ μH, $R_{Lmin} = 10$ Ω, $D_{nom} = 0.555$, $C = 51.2$ μF.

MOSFET: International Rectifier IRF142 n-channel power MOSFET with $V_{DSS} = 100$ V, $I_{SM} = 24$ A, $r_{DS} = 0.11$ Ω at $T = 25\,°$C, $C_o = 100$ pF, and $V_t = 4$ V.

Diode: ON Semiconductor MBR10100 with $V_{RRM} = 100$ V, $I_F = 10$ A, $R_F = 15$ mΩ, and $V_F = 0.8$ V.

Operational amplifier: Use LM741 or any ideal op-amp. Let the supply voltage for pulse-width modulator be $V_{CC} = 12$ V. The V_{EE} terminal can be connected to ground or zero potential.

Laboratory Manual for Pulse-Width Modulated DC–DC Power Converters, First Edition.
Marian K. Kazimierczuk and Agasthya Ayachit.
© 2016 John Wiley & Sons, Ltd. Published 2016 by John Wiley & Sons, Ltd.

Table 44.1   Parameters and their values

Parameter	Notation	Value
Nominal dc input voltage	V_I	28 V
DC output voltage	V_O	14 V
Switching frequency	f_s	100 kHz
Maximum output current	I_{Omax}	1.4 A
Nominal output current	I_{Onom}	1.0 A
Minimum output current	I_{Omin}	0.75 A
Output voltage ripple	V_r	$< 0.01V_O$

Feedback network, sawtooth generator, and pulse-width modulator: Pick $R_A = 4.65$ kΩ, $R_B = 5.35$ kΩ, and $R_C = 1$ kΩ. Let the value of the capacitor $C_c = 12$ nF. Use an ideal MOSFET in place of M_A. Let the reference voltage be equal to 1.4 V.

A. Simulation of the Buck Converter with Input Voltage Feedforward Control at Steady State

1. Construct the circuit shown in Figure 44.1 on the circuit simulator.
2. Connect a pulse voltage waveform at the gate-to-source terminals of the MOSFET M_A. Provide `initial voltage = 0 V`, `final/maximum voltage = 7 V`, `period = 10 μs`, and `ON time/width/duty cycle = 1 μs`. We only need to create periodic clock pulses to generate the gate-to-source voltage.
3. You may use an ideal transformer as an interface between the single-ended output terminal of the op-amp and the gate-to-source terminal of the MOSFET. The Spice code for the

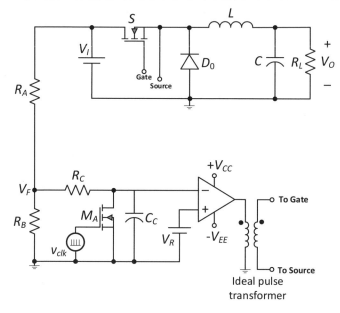

Figure 44.1   Circuit diagram of the PWM buck dc–dc converter with input voltage feedforward control.

ideal transformer as a subcircuit is given at the end of the handout. Alternatively, a dc/dc transformer with unity turns ratio can be used.

4. Enter the values of all the other components. Make sure the selection of MOSFET is appropriate and meets the desired ratings.
5. Set simulation type to `transient analysis`. Set end time = 20 ms and time step = 0.1 μs. Run the simulation.
6. Plot the following parameters after successful completion of the simulation. You may display the waveforms on different figure windows for better clarity.
   - Voltage at the inverting terminal of the op-amp, feedback voltage V_F, reference voltage V_R, and gate-to-source voltage v_{GS}.
   - Output voltage v_O, output current i_O, and output power p_O.
   - Inductor current i_L.
7. Change the value of the reference voltage should the output voltage be adjusted to obtain $V_O = 14$ V.

B. Simulation of the Buck Converter with Input Voltage Feedforward Control with Step Change in Input Voltage

1. Replace the input dc voltage source with a piecewise-linear voltage source as shown in Figure 44.2. Define the voltage profile as given in Table 44.2.
2. Set simulation type to `transient analysis`. Set end time = 20 ms and time step = 0.1 μs. Run the simulation.
3. Plot the following parameters after successful completion of the simulation. You may display the waveforms on different figure windows for better clarity.
   - Voltage at the inverting terminal of the op-amp, feedback voltage V_F, reference voltage V_R, and gate-to-source voltage v_{GS}.

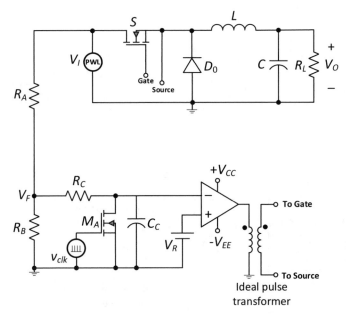

Figure 44.2   Circuit of the buck dc–dc converter with input voltage feedforward control used to simulate step change in the input voltage.

Table 44.2 Input voltage profile for the PWL voltage source

Time (ms)	0	5.9	6	13.9	14	16.9
Input voltage V_I (V)	28	28	34	14	30	30

- Output voltage v_O, output current i_O, and output power p_O.
- Inductor current i_L.

If the simulation tool permits, plot the waveform of duty cycle versus time to observe the feedforward action. Observe the transient characteristics of the above waveforms.

C. Simulation of the Closed-Loop Buck Converter for Step Change in the Load Resistance

1. Construct circuit shown in Figure 44.3 on the circuit simulator. The circuit consists of a series connection of a resistor R_x and an ideal MOSFET M and is placed in parallel with the load resistor R_L. Let $R_x = 10 \, \Omega$. This is needed in order to obtain the step change in the load resistance with respect to time.
2. Provide a piecewise-linear voltage to the gate of the MOSFET M. Enter the details in Table 44.3 in order to simulate a time-varying voltage to the gate of M.
3. Set simulation type to `transient analysis`. Set `end time = 20 ms` and `time step = 0.1 µs`. Run the simulation.
4. Plot the following parameters after successful completion of the simulation. You may display the waveforms on different figure windows for better clarity.

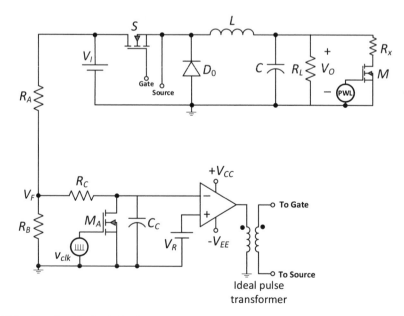

Figure 44.3 Circuit of the buck dc–dc converter with input voltage feedforward control used to simulate step change in the load resistance.

Table 44.3 Gate voltage profile for the PWL voltage source

Time (ms)	0	5.99	6	13.9	14
Gate voltage V_G (V)	0	0	10	10	0

- Voltage at the inverting terminal of the op-amp, feedback voltage V_F, reference voltage V_R, and gate-to-source voltage v_{GS}.
- Output voltage v_O, output current i_O, and output power p_O.
- Inductor current i_L.

5. Observe the transient characteristics of the above waveforms. Verify whether the output voltage is maintained at $V_O \approx 14$ V for the entire duration of time.

Post-lab Questions

1. What are the advantages of using feedforward control in dc–dc converters?
2. Explain the principle of operation of the buck converter with feedforward control in a few sentences.
3. How can the switching frequency be changed in the buck converter with feedforward control?
4. What are the basic differences between feedforward control and feedback control?

Part III

Semiconductor Materials and Power Devices

45

Temperature-Dependence of Si and SiC Semiconductor Materials

Objectives

The objectives of this lab are:

- To realize the differences in the physical properties of silicon (Si) and silicon-carbide (SiC) semiconductors.
- To analyze the variation in their characteristics with change in temperature.

Theory

The following analysis will show the dependence of the characteristics such as carrier concentration and resistivity of the two most commonly used semiconductor materials with respect to temperature.

The intrinsic carrier concentration n_i at any temperature T is given by

$$n_i = 2 \left(\frac{2\pi m_e kT}{h^2} \right)^{3/2} (k_e k_h)^{3/4} e^{-\frac{E_G}{2kT}}, \tag{45.1}$$

where $k = 8.617 \times 10^{-5}$ eV/K $= 1.3792 \times 10^{-23}$ J/K. Table H.2 in Appendix H provides the values of the essential physical properties of Si and SiC semiconductors.

The resistivity of the intrinsic semiconductors is expressed as

$$\rho_i = \frac{1}{q n_i (\mu_{n0} + \mu_{p0})}, \tag{45.2}$$

Laboratory Manual for Pulse-Width Modulated DC–DC Power Converters, First Edition.
Marian K. Kazimierczuk and Agasthya Ayachit.
© 2016 John Wiley & Sons, Ltd. Published 2016 by John Wiley & Sons, Ltd.

where

$$\mu_{n0} = \mu_{n(300)} \left(\frac{300}{T}\right)^{2.4},$$ (45.3)

and

$$\mu_{p0} = \mu_{p(300)} \left(\frac{300}{T}\right)^{2.2}.$$ (45.4)

The concentration of free electrons in the extrinsic n-type semiconductor is

$$n_n = \frac{N_D}{2}\left[\sqrt{1 + \left(\frac{2n_i}{N_D}\right)^2} + 1\right],$$ (45.5)

and the concentration of minority holes in the extrinsic n-type semiconductor is

$$p_n = \frac{N_D}{2}\left[\sqrt{1 + \left(\frac{2n_i}{N_D}\right)^2} - 1\right].$$ (45.6)

Procedure

A. Intrinsic Carrier Concentration Versus Temperature

1. Define all the variables on the MATLAB® editor window as given in the Table H.2 in Appendix H. Observe that the values of various properties are different for the two devices.
2. In this section, observe the effect of temperature on the intrinsic carrier concentration n_i as given in Equation (45.1) of the two semiconductors (Si and SiC). Vary the temperature from 200 to 1200 K in steps of 10 K.
3. Obtain the plots on the same figure window in order to compare their characteristics. Use MATLAB's® built-in semilogy command to plot the curves. Make sure your plots adhere to IEEE format.

B. Resistivity Versus Temperature

1. Refer to Equation (45.2) for the expression of resistivity. Define the equation on the MATLAB's® editor window.
2. Vary the temperature from 200 to 1200 K in steps of 10 K.
3. Sketch the plot of resistivity versus temperature of both the materials. All the curves must be plotted on the same figure window.

C. Extrinsic and Intrinsic Carrier Concentration Versus Temperature

1. Plot n_i, n_n, and p_n (in cm^{-3}) as a function of temperature for Si and SiC at $N_D = 10^{14}$ cm^{-3}. The equations for these variables are given in Equations (45.1), (45.5), (45.6), respectively. All the plots for the given value of N_D must be drawn on the same figure window.
2. Vary the temperature from 200 to 1200 K in steps of 10 K.
3. Repeat steps 1 and 2 in this section for $N_D = 10^{16}$ cm^{-3}.

Post-lab Questions

1. In Section A:
   - Define the term band gap energy. Comment on the difference in the values of the band gap energy of the three different semiconductors. You may support your answer using the electron–hole formation theory.
   - Which of the three semiconductor materials behave as an insulator at room temperature? Justify your answer.
2. In Section B:
   - Explain the significance of the plots in this section.
   - What happens to the resistivity of these semiconductor materials if you dope them with pentavalent or trivalent impurities?
3. In Section C:
   - Determine the maximum junction temperature of the two materials for all the cases in Section C. Define the term junction temperature.
   - What is the significance of these plots? Explain how the extrinsic and intrinsic carrier concentrations become equal at junction temperature.
   - Which of the two semiconductors, you think is suitable for high temperature applications?
   - What happens to the maximum junction temperature when the doping concentration is increased?

46

Dynamic Characteristics of the PN Junction Diode

Objectives

The objectives of this lab are:

- To compare the dynamic characteristics of silicon and silicon-carbide pn junction diodes at different temperatures.
- To observe the dependence of diode characteristics on temperature using a circuit simulator.
- To observe the rectification properties of a power diode at different operating temperatures.

Theory

The Shockley diode equation for the forward and the reverse-biased regions is

$$I_D = I_S \left(e^{\frac{V_D}{nV_T}} - 1 \right), \tag{46.1}$$

where $n = 2$ is the emission coefficient and I_S is the reverse saturation current and given by

$$I_S(T) = A_J q \left(\frac{D_p}{L_p} p_{n0} + \frac{D_n}{L_n} n_{p0} \right). \tag{46.2}$$

In the above expression, A_J represents the cross-sectional area of the pn junction, $p_{n0} = \frac{n_i^2}{N_D}$ is the concentration of holes in the n region, $n_{p0} = \frac{n_i^2}{N_A}$ is the concentration of electrons in the p region, $D_p = \mu_{p0} V_T = \frac{\mu_{p0} kT}{q}$ is the diffusion coefficient of holes, $D_n = \mu_{n0} V_T = \frac{\mu_{n0} kT}{q}$ is the diffusion coefficient of electrons, $L_n = \sqrt{D_n \tau_n}$ and $L_p = \sqrt{D_p \tau_p}$ are the diffusion lengths

Laboratory Manual for Pulse-Width Modulated DC–DC Power Converters, First Edition.
Marian K. Kazimierczuk and Agasthya Ayachit.
© 2016 John Wiley & Sons, Ltd. Published 2016 by John Wiley & Sons, Ltd.

of electrons and holes, respectively, and τ_n and τ_p are the average lifetimes of the minority electrons and holes, respectively.

The expression for the intrinsic carrier concentration n_i can be obtained from Equation (45.1) in Lab 45.

Specifications

$A_J = 1\text{cm}^2$, $N_A = 10^{16}\text{cm}^{-3}$, $N_D = 10^{14}\text{cm}^{-3}$.

Assume: $\tau_n = 10$ µs. Estimate τ_p using $\tau_n/\tau_p = N_A/N_D$.

The values of the other physical properties can be obtained from Table H.2 in Appendix H.

Procedure
A. Dynamic Characteristics of Silicon PN Junction Diodes

1. Define all the variables on the MATLAB® editor window. The physical constants and properties of the silicon semiconductor material can be found in Appendix H.
2. Sketch the plot of the diode current I_D as given in Equation (46.1) as a function of the diode voltage V_D for $T = 300$ K, 350 K, and 400 K. Use Equation (46.2) in order to define the reverse saturation current I_S.
3. Define a variable V_D. Vary V_D from 0 to 1 V in steps of 0.001 V.
4. Limit the range of y-axis to 0.4 mA in order to make clear distinction between the plots. Make sure to express the current on the y-axis in terms of milliamperes.

B. Dynamic Characteristics of Silicon-Carbide PN Junction Diodes

1. Define all the variables on the MATLAB® editor window. The various parameters of the silicon-carbide semiconductor are provided in Appendix H.
2. Sketch the plot of the diode current I_D as given in Equation (46.1) as a function of the diode voltage V_D for $T = 300$ K, 423 K, and 573 K. Use Equation (46.2) in order to represent the reverse saturation current I_S.
3. Define a variable V_D. Vary V_D from 2 to 6 V in steps of 0.001 V.
4. Limit the range of y-axis to 0.4 mA in order to make clear distinction between the plots. Make sure to express the current on the y-axis in terms of milliamperes.

C. Determination of Threshold Voltage of a Power Diode using Circuit Simulator

1. Construct the circuit as shown in Figure 46.1 on the circuit simulator. Use the MUR1560 power diode available in the component library. Let $R_D = 10 \ \Omega$. For readers using PSpice, the spice model of MUR1560 power diode is provided in Appendix F.
2. For the dc voltage source assign an arbitrary dc voltage value (e.g., 1 V).
3. Perform dc analysis. Sweep the value of the dc voltage source from 0 to 1 V in steps of 0.001 V.

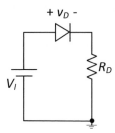

Figure 46.1   Circuit diagram to simulate the diode characteristics.

4. Plot the waveform of the current flowing from the anode to cathode of the diode. Determine the value of the threshold voltage of the diode. This value corresponds to the threshold voltage at ambient temperature ($T = 25°C$).

5. Next, analyze the circuit at different temperatures. Change the the circuit temperature to 75°C and then to 120°C.

   NOTE On spice-based simulators, enter the command .TEMP 27 75 120. On SABER circuit simulator, double-click on the resistor and the diode. In the property editor window, change the value of the temperature parameter temp to 75. Simulate the circuit and plot the waveform of the diode current. Make sure to append the plots in the output figure window. Resimulate the circuit by changing temp to 120.

Post-lab Questions

1. Explain in a few sentences the development and operation of the pn junction diode. Cite references appropriately.

2. In Section A,
   - Explain the significance of the reverse saturation current I_S. Indicate whether I_S is temperature dependent.
   - What are the threshold voltages V_{th} of the three $i_D - v_D$ characteristics? Explain how the threshold voltages vary with temperature.

3. In Section B,
   - What are the threshold voltages V_{th} of the three $i_D - v_D$ characteristics? Explain how the threshold voltage vary with operating temperature.
   - Explain how the SiC device is capable of withstanding high temperatures, when compared with its Si counterpart.
   - Also explain why the threshold voltage of the SiC device is higher than that of Si.

4. In Section C,
   - Identify the threshold voltage for the three different temperatures.
   - Obtain the Spice model of MUR1560 or its data sheet and identify the maximum junction temperature or the operating temperature.

47

Characteristics of the Silicon and Silicon-Carbide PN Junction Diodes

Objectives

The objectives of this lab are:

- To observe the variation in the width of the depletion layer for variation in the applied diode voltage.
- To observe the variation in the breakdown voltage for different doping concentrations.
- To observe the variation in the nonideal parasitic junction capacitance in the pn junction diode as a function of the applied voltage.

Theory

The intrinsic carrier concentration n_i at any temperature T is given by

$$n_i = 2 \left(\frac{2\pi m_e kT}{h^2} \right)^{3/2} \left(k_e k_h \right)^{3/4} e^{-\frac{E_G}{2kT}}, \tag{47.1}$$

where $k = 8.617 \times 10^{-5} \text{eV/K} = 1.3792 \times 10^{-23} \text{J/K}$. The Table H.2 in Appendix H provides the values of the essential physical properties of Si and SiC materials.

The width of the depletion region is expressed as

$$W = \sqrt{\frac{2\epsilon_0 \epsilon_r (V_{bi} - v_D)}{q} \left(\frac{1}{N_A} + \frac{1}{N_D} \right)}, \tag{47.2}$$

Laboratory Manual for Pulse-Width Modulated DC–DC Power Converters, First Edition.
Marian K. Kazimierczuk and Agasthya Ayachit.
© 2016 John Wiley & Sons, Ltd. Published 2016 by John Wiley & Sons, Ltd.

where ϵ_r is the relative permittivity of the semiconductor material and V_{bi} is the built-in potential expressed as

$$V_{bi} = \frac{kT}{q} \ln \left(\frac{N_A N_D}{n_i^2} \right). \tag{47.3}$$

The breakdown voltage due to the avalanche breakdown process is expressed as

$$V_{BD} = \frac{\epsilon_0 \epsilon_r \left(1 + \frac{N_D}{N_A} \right) E_{BD}^2}{2 q N_D} - V_{bi}. \tag{47.4}$$

The junction capacitance of the pn junction diode is

$$C_J = \frac{C_{J0}}{\left(1 - \frac{v_D}{V_{bi}} \right)^m}, \tag{47.5}$$

where m is the gradient coefficient and $m = 0.5$ for the step junction, and C_{J0} is the junction capacitance at $v_D = 0$ V expressed as

$$C_{J0} = A_J \sqrt{\frac{q \epsilon_0 \epsilon_r}{2 V_{bi} \left(\frac{1}{N_A} + \frac{1}{N_D} \right)}}. \tag{47.6}$$

Specifications

$A_J = 4$ mm^2, $N_A = 10^{16}$ cm^{-3}

The values of the physical constants and semiconductor material properties can be obtained from Table H.2 in Appendix H.

Procedure

A. Analysis of Depletion-Layer Width as a Function of Diode Voltages of Si and SiC

1. Define all the variables on the MATLAB® editor window. The various parameters of the silicon semiconductor are provided in Appendix H.
2. Plot the width W as given in Equation (47.2) as a function of the diode voltage v_D for $N_D = 10^{14}$cm^{-3}, $N_D = 10^{15}$cm^{-3}, and $N_D = 10^{16}$cm^{-3}. Vary v_D from 0 to -600 V in steps of 1 V.
3. On a new figure, repeat step (1) for the silicon-carbide pn junction. Use the same doping concentrations.
4. Make sure to express the current on the y-axis in terms of microamperes.

B. Analysis of the Breakdown Voltage as a Function of Doping Concentration for Si, SiC, and GaN

1. Define all the variables on the MATLAB® editor window.
2. Plot the breakdown voltage V_{BD} as given in Equation (47.4) as a function of the doping concentration for silicon and silicon carbide at $T = 300$ K.
3. Vary N_D from 10^{12} to 10^{17} cm^{-3} in steps of 10^{12} cm^{-3}.

C. Analysis of Junction Capacitance as a Function of Applied Voltage for Si, SiC, and GaN

1. Define all the variables on the MATLAB® editor window.
2. Plot the junction capacitances as given in Equation (47.5) as a function of the diode voltage at $T = 300$ K for the silicon pn junction diode. Let $m = 0.5$.
3. Vary v_D from 0 to -20 V in steps of 0.1 V.
4. Using the same details as above, plot the junction capacitance as a function of the diode voltage for the silicon-carbide pn junction diode.

Post-lab Questions

1. In Section A:
   - Define the term depletion-layer width. How does the depletion-layer width vary with the applied diode voltage?
   - Estimate the value of the zero-bias depletion region width.
   - How does the doping concentration affect the width of the n and p region in the diode? What is the maximum width of the depletion region that can be achieved, theoretically?
2. In Section B:
   - Explain in a few sentences the avalanche breakdown mechanism.
   - How does the doping concentration N_D affect the breakdown voltage?
   - Derive an analytical approximate expression to calculate the maximum value of N_D?
3. In Section C:
   - Explain in a few sentences the formation of junction capacitance.
   - Explain how C_{J0} varies as the doping concentration N_D is increased.

48

Analysis of the Output and Switching Characteristics of Power MOSFETs

Objectives

The objectives of this lab are:

- To analyze the transfer characteristics of power MOSFETs on a circuit simulator.
- To analyze the switching characteristics of power MOSFETs on a circuit simulator.

Procedure

A. Analysis of i_D Versus v_{GS} Characteristics of Power MOSFET

1. Construct the circuit as shown in Figure 48.1 on the circuit simulator. Use a drain resistance R_D of 1 Ω. Use IRF540 n-channel power MOSFET. Provide $V_{GS} = 5$ V and $V_{DD} = 10$ V.
2. In the simulation options, select DC Sweep (Transfer) Analysis.
3. In this analysis, V_{GS} represents the independent voltage source, whose value is to be varied. Select the appropriate source reference from the drop-down menu option.
4. Let the voltage vary from 0 to 10 V in steps of 0.1 V.
5. Run the simulation and plot the drain current.
6. Next, change the value of the supply voltage V_{DD} to 15 V on the circuit schematic. Observe the drain current waveform. Repeat the above steps for $V_{DD} = 20$ V and $V_{DD} = 25$ V. Append the plots on the same figure window for comparison.

B. Analysis of i_D Versus v_{DS} Characteristics of Power MOSFET

1. Consider the circuit shown in Figure 48.1. Provide $V_{GS} = 5$ V and $V_{DD} = 10$ V.

Laboratory Manual for Pulse-Width Modulated DC–DC Power Converters, First Edition.
Marian K. Kazimierczuk and Agasthya Ayachit.
© 2016 John Wiley & Sons, Ltd. Published 2016 by John Wiley & Sons, Ltd.

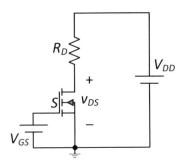

Figure 48.1　Circuit schematic to determine the electrical characteristics of the MOSFET.

2. In this section, the value of the voltage V_{DS} is to be varied and the drain current is to be plotted for different V_{GS} values. Use a drain resistance R_D of 1 Ω.
3. Vary the voltage supply V_{DD} from 0 to 75 V in steps of 0.1 V.
4. Run the simulation and plot the drain current.
5. Next, change the voltage V_{GS} to 10 V. Observe the drain current waveform. Repeat the steps for $V_{GS} = 15$ V and $V_{GS} = 20$ V. Append the plots on the same figure window for comparison.

C. Analysis of Switching Characteristics

1. Replace the dc voltage source V_{GS} with a pulse voltage source. Provide a pulse voltage source with a frequency 100 kHz and a duty ratio of 0.5. Let the maximum amplitude be 1 V. Let $V_{DD} = 12$ V. Use a drain resistance of 1 Ω.
2. Plot the gate-to-source voltage waveform, drain-to-source voltage waveform, and the drain current waveform.
3. Increase the switching frequency to 1 MHz and observe the waveforms.
4. Repeat the experiment for 5 MHz, 10 MHz, and 20 MHz.

Post-lab Questions

1. From plots obtained in Section A, identify the value of threshold voltage V_T. Obtain the data sheet for IRF540. Compare the values of gate-to-source threshold voltage V_T provided in the data sheet with that obtained through simulations. How do they compare?
2. From plots obtained in Section C, observe the voltage and current waveforms for different frequencies mentioned. What happens to the waveforms at higher operating frequencies? Check whether the MOSFET behaves as a switch at higher switching frequencies. If not, then explain why?
3. For low-frequency operation, observe the Miller Plateau effect. Explain the concept of Miller Plateau and Miller Effect in a few sentences.

49

Short-Channel Effects in MOSFETs

Objectives

The objectives of this lab are:

- To observe the variation in the average electron drift velocity with respect to electric field.
- To observe the variation in the electron mobility with respect to electric field.
- To observe the variation in the mobility with respect to temperature and electric field.
- To understand the differences between linear law, square law, and the exact drain current equation.
- To observe the impact of electric field on the drain current of the MOSFET.

Theory

The average drift electron velocity in terms of the electric field intensity is given as

$$v_n = \frac{\mu_{n0}E}{1 + \frac{\mu_{n0}E}{v_{sat}}}, \tag{49.1}$$

where μ_{n0} is the low-field electron mobility and v_{sat} is the electron drift saturation velocity. The electron field-dependent mobility is given by

$$\mu_n = \frac{v_n}{E}. \tag{49.2}$$

The relationship between the electron field-dependent mobility and temperature is given by

$$\mu_n = \frac{\mu_{n0}E}{1 + \frac{\mu_{n0}E}{v_{sat}}} \left(\frac{300}{T}\right)^{2.4}. \tag{49.3}$$

Laboratory Manual for Pulse-Width Modulated DC–DC Power Converters, First Edition.
Marian K. Kazimierczuk and Agasthya Ayachit.
© 2016 John Wiley & Sons, Ltd. Published 2016 by John Wiley & Sons, Ltd.

The equation of the drain current determined by the *square law* is given by

$$i_D = \frac{1}{2}\mu_{n0}C_{ox}\left(\frac{W}{L}\right)(v_{GS} - V_t)^2(1 + \lambda v_{DS}),\tag{49.4}$$

where C_{ox} is the gate-oxide capacitance, λ is the channel-length modulation factor, and W/L is the aspect ratio of the channel of the MOSFET.

The equations of the drain current due to short channel determined by the *linear law* is

$$i_D = \frac{1}{2}\frac{\mu_{n0}}{1 + \theta(v_{GS} - V_t)}C_{ox}\left(\frac{W}{L}\right)(v_{GS} - V_t)^2(1 + \lambda v_{DS}),\tag{49.5}$$

where θ is the mobility degrading coefficient.

The equations of the drain current due to short channel determined by the *exact equation* is

$$i_D = \frac{1}{2}C_{ox}WV_{sat}(v_{GS} - V_t)(1 + \lambda v_{DS}).\tag{49.6}$$

The equation of the drain current in the ohmic region as a function of the electric field is expressed as

$$i_D = \frac{\mu_{n0}}{1 + \frac{\mu_{n0}E}{v_{sat}}}C_{ox}\left(\frac{W}{L}\right)\left[(V_{GS} - V_t)v_{DS} - \frac{v_{DS}^2}{2}\right].\tag{49.7}$$

The equation of the drain current in the pinch-off region as a function of the electric field is expressed as

$$i_D = \frac{1}{2}\frac{\mu_{n0}}{1 + \frac{\mu_{n0}E}{v_{sat}}}C_{ox}\left(\frac{W}{L}\right)v_{DS}^2(1 + \lambda v_{DS}).\tag{49.8}$$

Specifications

The values of the physical properties are provided in Table 49.1.

Table 49.1　Physical properties and their values

Parameter	Notation	Value
Low-field mobility	μ_{n0}	1360 cm^2/V · s
Electron drift saturation velocity	v_{sat}	8×10^6 cm/s
Aspect ratio	W/L	10^5 for $L = 0.5$ μm
Gate-oxide capacitance	C_{ox}	0.33 mF/m^2
Gate-to-source threshold voltage	V_t	0.356 V
Channel-length modulation factor	λ	0.051/V
Mobility degrading coefficient	θ	1.5

Procedure

A. Velocity and Electron Mobility as a function of Electric Field

1. Define Equations (49.1) and (49.2) on MATLAB®. Use the values of the parameters provided in Table 49.1. Make sure the units are mentioned appropriately (either in centimeter or meter).
2. Vary E from 10 to 10^7 V/cm in steps of 100 V/cm. Obtain the plots of the average electron drift velocity v_n and the electron mobility μ_n as a function of electric field E.
3. The two plots must be on different figure windows. Make sure your plots are in IEEE format.

B. Electron Mobility as a Function of Electric Field and Temperature

1. Define the temperature T as a variable such that it varies from 300 to 500 K in steps of 10 K.
2. Define the `meshgrid` to plot a 3D figure of mobility μ_n as given in Equation (49.3) as a function of the electric field E and temperature T.
3. The x-axis must represent the electric field E varying from 10 to 10^7 V/cm, the y-axis must represent the temperature T for the values mentioned above, and the z-axis must represent the mobility μ_n. All the plots must be in IEEE format.

C. Variation in Drain Current for Change in the Gate-to-Source Voltage for the Different Laws

1. The expressions for drain current determined by the square law, exact equation, and linear law are given in Equations (49.4), (49.6), and (49.5).
2. Define the equations on the MATLAB® editor window. Let $v_{DS} = V_{DS} = 10$ V.
3. Plot the drain current i_D with respect to variation in the gate-to-source voltage v_{GS} for the three laws on the same figure. Vary v_{GS} from 0.5 to 3 V in steps of 0.001 V.
4. Mention clearly on the plot the laws which they represent either using text command or legend command.

Post-lab Questions

1. Provide your conclusions for the plots obtained in Sections A, B, and C.
2. What are the differences between the three different equations for the drain current through a MOSFET?
3. Explain the term *channel-length modulation*.
4. What is *mobility degradation coefficient*?

50

Gallium-Nitride Semiconductor: Material Properties

Objectives

The objectives of this lab are as follows:

- To determine the dependence of the intrinsic carrier concentration of gallium-nitride semiconductors on temperature.
- To observe the variation in the mobility of carriers with changes in temperatures and carrier concentration.

Theory

The intrinsic carrier concentration n_i at any temperature T is given by

$$n_i = 2 \left(\frac{2\pi m_e kT}{h^2} \right)^{3/2} \left(k_e k_h \right)^{3/4} e^{-\frac{E_G}{2kT}}, \tag{50.1}$$

where $k = 8.617 \times 10^{-5}$ eV/K $= 1.3792 \times 10^{-23}$ J/K. Table H.2 in Appendix H provides the values of all essential physical properties of the GaN semiconductor material.

The low-field electron mobility μ_e as a function of temperature T and carrier concentration N_D is given by

$$\mu_e = \mu_{min} + \frac{\mu_L - \mu_{min}}{1 + \left(\frac{N_D}{N_{D0}} \right)^{\gamma_0}}, \tag{50.2}$$

Laboratory Manual for Pulse-Width Modulated DC–DC Power Converters, First Edition.
Marian K. Kazimierczuk and Agasthya Ayachit.
© 2016 John Wiley & Sons, Ltd. Published 2016 by John Wiley & Sons, Ltd.

where

$$\mu_{min} = \mu_{min(300)} \left(\frac{T}{300} \right)^{\gamma_1},$$

$$\mu_L = \mu_{L(300)} \left(\frac{T}{300} \right)^{\gamma_1},$$

and

$$N_{D0} = N_{D(300)} \left(\frac{T}{300} \right)^{\gamma_2}.$$

In the above expressions, $\mu_{L(300)}$ represents the maximum value of mobility at $T = 300$ K, $\mu_{min(300)}$ represents the lowest value of mobility at $T = 300$ K, $N_{D(300)}$ is the reference carrier concentration at $T = 300$ K, N_c is the carrier concentration normalized with respect to N_{D0}, and $\gamma_0, \gamma_1, \gamma_2$ are the empirical constants.

The resistivity of the intrinsic GaN semiconductor is expressed as

$$\rho_i = \frac{1}{qn_i\mu_e}. \tag{50.5}$$

Procedure

A. Intrinsic Carrier Concentration Versus Temperature

1. Define all the constants on the MATLAB® editor window as given in the Table H.2 in Appendix H.
2. In this section, observe the effect of temperature T on the intrinsic carrier concentration n_i as given in Equation (50.1) for the GaN semiconductor. Vary the temperature from 200 to 1200 K in steps of 10 K.
3. Use MATLAB's® built-in semilogy command to plot the curve. Make sure your plots adhere to IEEE format.

B. Mobility Versus Temperature

1. Refer to Equation (50.2) for the expression of mobility. Define the equation on the MATLAB's® editor window. Use the following values for the constant terms:
$N_{D0} = 10^{17}\,\text{cm}^{-3}, N_D = 10^{16}\,\text{cm}^{-3}, \mu_{L(300)} = 1600\,\text{cm}^2/\text{V} \cdot \text{s}, \mu_{min(300)} = 100\,\text{cm}^2/\text{V} \cdot \text{s},$
$\gamma_0 = 0.3, \gamma_1 = -3, \gamma_0 = 4.4.$
2. Vary the temperature from 200 to 1200 K in steps of 10 K.
3. Sketch the plot of mobility versus temperature. Use MATLAB's® built-in plot command to plot the curve.

C. Resistivity Versus Temperature

1. Refer to Equation (50.5) for the expression of resistivity. Define the equation on the MATLAB's® editor window. The expressions for the intrinsic carrier concentration and the mobility are provided in Equations (50.1) and (50.2).

2. Vary the temperature from 200 to 1200 K in steps of 10 K.
3. Sketch the plot of resistivity versus temperature. Use MATLAB's® built-in semilogy command to plot the curve.

D. Mobility Versus Donor Concentration

1. Refer to Equation (50.2) for the expression of mobility.
2. Vary the value of N_D from 10^{10} to 10^{20} cm^{-3}. Let:
   $N_{D0} = 10^{17}$ cm^{-3}, $T = 300$ K, $\mu_{L(300)} = 1600$ cm^2/V $\cdot$ s, $\mu_{min(300)} = 100$ cm^2/V $\cdot$ s, $\gamma_0 = 0.3$, $\gamma_1 = -3$, $\gamma_0 = 4.4$.
3. Use the function `logspace` command to define the variable. The syntax is:

   ```
   ND = logspace(10, 20, 1e3).
   ```

4. Sketch the plot of mobility versus the donor concentration. Use the command `semilogx` to plot the curve. Label the axes' details clearly.

Post-lab Question

Explain the significance of all the plots obtained in Sections A, B, C, and D.

Appendices

A

Design Equations for Continuous-Conduction Mode

Common Equations Needed for the Design of Converters

DC Output Power

$$P_{Omin} = V_O I_{Omin} = \frac{V_O^2}{R_{Lmax}}. \tag{A.1}$$

$$P_{Omax} = V_O I_{Omax} = \frac{V_O^2}{R_{Lmin}}. \tag{A.2}$$

DC Voltage Transfer Function

$$M_{V\,DCmin} = \frac{V_O}{V_{Imax}}. \tag{A.3}$$

$$M_{V\,DCnom} = \frac{V_O}{V_{Inom}}. \tag{A.4}$$

$$M_{V\,DCmax} = \frac{V_O}{V_{Imin}}. \tag{A.5}$$

Specific Expressions for the Design of Converters in CCM

The equations for estimating the values of the different parameters and components of the buck, boost, and buck–boost dc–dc converters in continuous-conduction mode (CCM) are provided in Table A.1.

Similarly, the equations for estimating the values of different parameters and components for the single-output flyback and the forward converters in CCM are provided in Tables A.2 and A.3.

Laboratory Manual for Pulse-Width Modulated DC–DC Power Converters, First Edition.
Marian K. Kazimierczuk and Agasthya Ayachit.
© 2016 John Wiley & Sons, Ltd. Published 2016 by John Wiley & Sons, Ltd.

Table A.1 Steady-state design equations for buck, boost, and buck–boost dc–dc converters in CCM

Parameter/topology	Buck	Boost	Buck–boost
Min. duty cycle D_{min}	$\dfrac{M_{VDCmin}}{\eta}$	$1-\dfrac{\eta}{M_{VDCmin}}$	$\dfrac{M_{VDCmin}}{M_{VDCmin}+\eta}$
Nom. duty cycle D_{nom}	$\dfrac{M_{VDCnom}}{\eta}$	$1-\dfrac{\eta}{M_{VDCnom}}$	$\dfrac{M_{VDCnom}}{M_{VDCnom}+\eta}$
Max. duty cycle D_{max}	$\dfrac{M_{VDCmax}}{\eta}$	$1-\dfrac{\eta}{M_{VDCmax}}$	$\dfrac{M_{VDCmax}}{M_{VDCmax}+\eta}$
Min. inductance L_{min}	$\dfrac{R_{Lmax}(1-D_{min})}{2f_s}$	$\dfrac{2}{27}\dfrac{R_{Lmax}}{f_s}$ for $D_{min}<1/3$ $\dfrac{R_{Lmax}D_{min}(1-D_{min})^2}{2f_s}$ for $D_{min}>1/3$	$\dfrac{R_{Lmax}(1-D_{min})^2}{2f_s}$
Max. inductor current ripple Δi_{Lmax} (where $L>L_{min}$)	$\dfrac{V_O(1-D_{min})}{f_s L}$	$\dfrac{V_O D_{min}(1-D_{min})}{f_s L}$	$\dfrac{V_O(1-D_{max})}{2f_s}$
Max. ESR of filter capacitor r_{Cmax}	$\dfrac{V_r}{\Delta i_{Lmax}}$	$\dfrac{V_r}{2I_{DMmax}}$	$\dfrac{V_r}{2I_{DMmax}}$
Min. filter capacitance C_{min} (where $r_C<r_{Cmax}$)	$\max\left[\dfrac{D_{max}}{2f_s r_C},\dfrac{1-D_{min}}{2f_s r_C}\right]$	$\dfrac{2D_{max}}{f_s R_{Lmin}}\dfrac{V_O}{V_r}$	$\dfrac{2D_{max}}{f_s R_{Lmin}}\dfrac{V_O}{V_r}$
Max. MOSFET voltage stress V_{SM}	V_{Imax}	V_O	V_O+V_{Imax}
Max. diode voltage stress V_{DM}	V_{Imax}	V_O	V_O+V_{Imax}
Max. MOSFET current stress I_{SM}	$I_{Omax}+\dfrac{\Delta i_{Lmax}}{2}$	$\dfrac{I_{Omax}}{1-D_{max}}+\dfrac{V_O D_{max}(1-D_{max})}{2f_s L}$	$I_{Imax}+I_{Omax}+\dfrac{\Delta i_{Lmax}}{2}$
Max. diode current stress I_{DM}	$I_{Omax}+\dfrac{\Delta i_{Lmax}}{2}$	$\dfrac{I_{Omax}}{1-D_{max}}+\dfrac{V_O D_{max}(1-D_{max})}{2f_s L}$	$I_{Imax}+I_{Omax}+\dfrac{\Delta i_{Lmax}}{2}$

Table A.2 Steady-state design equations for the flyback and forward converters in CCM

Parameter/topology	Flyback	Forward
Turns ratio n for assumed D_{max}	$\dfrac{\eta D_{max}}{(1 - D_{max}M_{VDCmax})}$	$n_1 = n_3 = \dfrac{\eta D_{max}}{M_{VDCmax}}$ [a]
Min. duty cycle D_{min}	$\dfrac{nM_{VDCmin}}{nM_{VDCmin} + \eta}$	$\dfrac{n_1 M_{VDCmin}}{\eta}$
Nom. duty cycle D_{nom}	$\dfrac{nM_{VDCnom}}{nM_{VDCnom} + \eta}$	$\dfrac{n_1 M_{VDCnom}}{\eta}$
Max. duty cycle D_{max}	$\dfrac{nM_{VDCmax}}{nM_{VDCmax} + \eta}$	$\dfrac{n_1 M_{VDCmax}}{\eta}$
Min. magnetizing inductance $L_{m(min)}$	$\dfrac{n^2 R_{Lmax}(1 - D_{min})^2}{2f_s}$	$\dfrac{D_{min} V_{Imax}}{f_s \Delta i_{Lm(max)}}$
Min. inductance L_{min}	-not applicable-	$\dfrac{R_{Lmax}(1 - D_{min})}{2f_s}$
Max. current ripple through the magnetizing inductance $\Delta i_{Lm(max)}$ (where $L_m > L_{m(min)}$)	$\dfrac{nV_O(1 - D_{min})}{f_s L_m}$	$\alpha \dfrac{I_{Omax} + \dfrac{\Delta i_{Lm(max)}}{2}}{n_1}$ [b]
Max. current ripple through output filter inductance Δi_{Lmax} (where $L > L_{min}$)	-not applicable-	$\dfrac{V_O(1 - D_{min})}{f_s L}$

[a] where n_1 is the turns ratio between the primary and secondary winding and n_2 is the turns ratio between primary winding and tertiary (reset) winding.

[b] where α is a constant number less than 1.

Table A.3 Steady-state design equations for the flyback and forward converters in CCM

Parameter / Topology	Flyback	Forward
Max. ESR of filter capacitor r_{Cmax}	$\dfrac{V_r}{2I_{DMmax}}$	$\dfrac{V_r}{\Delta i_{Lmax}}$
Min. filter capacitance C_{min} (where $r_C < r_{Cmax}$)	$\dfrac{2D_{max}V_O}{f_s R_{Lmin} V_r}$	$\max\left[\dfrac{D_{max}}{2f_s r_C}, \dfrac{1-D_{min}}{2f_s r_C}\right]$
Max. MOSFET voltage stress V_{SM}	$V_{Imax} + nV_O$	$\left(\dfrac{n_1}{n_3}+1\right)V_{Imax}$
Max. diode voltage stress V_{DM}	$\dfrac{V_{Imax}}{n}+V_O$	$V_{D1max} = \dfrac{V_{Imax}}{n_3}$
		$V_{D2max} = \dfrac{V_{Imax}}{n_1}$
		$V_{D3max} = \left(\dfrac{n_3}{n_1}+1\right)V_{Imax}$
Max. MOSFET current stress I_{SM}	$I_{Imax}+\dfrac{I_{Omax}}{n}+\dfrac{\Delta i_{Lmax}}{2}$	$I_{Imax}+\Delta i_{Lm(max)}$
Max. diode current stress I_{DM}	$nI_{Imax}+I_{Omax}+\dfrac{n\Delta i_{Lmax}}{2}$	$I_{D1max}=I_{D2max}=I_{Omax}+\dfrac{\Delta i_{Lmax}}{2}$
		$I_{D3max}=\left(\dfrac{n_1}{n_3}\right)\Delta i_{Lm(max)}$

B

Design Equations for Discontinuous-Conduction Mode

Specific Expressions for the Design of Converters in DCM

The equations for estimating the values of the different parameters and components of the buck, boost, and buck–boost dc–dc converters in discontinuous-conduction mode (DCM) are provided in Table B.1.

Similarly, the equations for estimating the values of different parameters and components for the single-output flyback and the forward converters in DCM are provided in Tables B.2 and B.3.

Laboratory Manual for Pulse-Width Modulated DC–DC Power Converters, First Edition.
Marian K. Kazimierczuk and Agasthya Ayachit.
© 2016 John Wiley & Sons, Ltd. Published 2016 by John Wiley & Sons, Ltd.

Table B.1 Steady-state design equations for buck, boost, and buck–boost dc–dc converters in DCM

Parameter/topology	Buck	Boost	Buck–boost
Max. inductance L_{max} (for D_{Bmax} assumed)	$\dfrac{R_{Lmin}(1-D_{Bmax})}{2f_s}$	$\dfrac{R_{Lmin}D_{Bmin}(1-D_{Bmin})^2}{2f_s}$ for $D_{min}<1/3$ $\dfrac{R_{Lmin}D_{Bmax}(1-D_{Bmax})^2}{2f_s}$ for $D_{min}>1/3$	$\dfrac{R_{Lmin}(1-D_{Bmax})^2}{2f_s}$
Min. duty cycle D_{min}	$\sqrt{\dfrac{2f_sLM_{VDCmin}^2}{\eta R_{Lmin}(1-M_{VDCmin})}}$	$\sqrt{\dfrac{2f_sLM_{VDCmin}(1-M_{VDCmin})}{\eta R_{Lmin}}}$	$M_{VDCmin}\sqrt{\dfrac{2f_sL}{\eta R_{Lmin}}}$
Nom. duty cycle D_{nom}	$\sqrt{\dfrac{2f_sLM_{VDCnom}^2}{\eta R_{Lmin}(1-M_{VDCnom})}}$	$\sqrt{\dfrac{2f_sLM_{VDCnom}(1-M_{VDCnom})}{\eta R_{Lmin}}}$	$M_{VDCnom}\sqrt{\dfrac{2f_sL}{\eta R_{Lmin}}}$
Max. duty cycle D_{max}	$\sqrt{\dfrac{2f_sLM_{VDCmax}^2}{\eta R_{Lmin}(1-M_{VDCmax})}}$	$\sqrt{\dfrac{2f_sLM_{VDCmax}(1-M_{VDCmax})}{\eta R_{Lmin}}}$	$M_{VDCmax}\sqrt{\dfrac{2f_sL}{\eta R_{Lmin}}}$
Max. inductor current ripple Δi_{Lmax} (where $L<L_{max}$)	$\dfrac{D_{min}(V_{Imax}-V_O)}{f_sL}$	$\dfrac{D_{max}V_{Imin}}{f_sL}$	$\dfrac{V_{Imax}D_{min}}{f_sL}$
Max. ESR of filter capacitor r_{Cmax}	$\dfrac{V_r}{\Delta i_{Lmax}}$	$\dfrac{V_r}{2\Delta i_{Lmax}}$	$\dfrac{V_r}{\Delta i_{Lmax}}$
Min. filter capacitance C_{min} (where $r_C<r_{Cmax}$)	$\dfrac{1}{2r_Cf_s}$	$\dfrac{2D_{max}V_O}{f_sR_{Lmin}V_r}$	$\dfrac{D_{max}}{f_sR_{Lmin}}\dfrac{V_O}{0.25V_r}$
Max. MOSFET voltage stress V_{SM}	V_{Imax}	V_O	V_O+V_{Imax}
Max. diode voltage stress V_{DM}	V_{Imax}	V_O	V_O+V_{Imax}
Max. MOSFET current stress I_{SM}	Δi_{Lmax}	Δi_{Lmax}	Δi_{Lmax}
Max. diode current stress I_{DM}	Δi_{Lmax}	Δi_{Lmax}	Δi_{Lmax}

Table B.2 Steady-state design equations for the flyback and forward dc–dc converters in DCM

Parameter/topology	Flyback	Forward
Turns ratio n for assumed D_{Bmax}	$\dfrac{\eta D_{Bmax}}{(1 - D_{Bmax}M_{VDCmax})}$	$n_1 = n_3 = \dfrac{\eta D_{Bmax}}{M_{VDCmax}}$ [a]
Min. duty cycle D_{min} for $L_m < L_{m(max)}$	$M_{VDCmin}\sqrt{\dfrac{2f_sL_m}{\eta R_{Lmin}}}$	$\sqrt{\dfrac{2f_sLn_1^2M_{VDCmin}^2}{\eta(1 - n_1M_{VDCmin})R_{Lmin}}}$
Nom. duty cycle D_{nom}	$M_{VDCnom}\sqrt{\dfrac{2f_sL_m}{\eta R_{Lmin}}}$	$\sqrt{\dfrac{2f_sLn_1^2M_{VDCnom}^2}{\eta(1 - n_1M_{VDCnom})R_{Lmin}}}$
Max. duty cycle D_{max}	$M_{VDCmax}\sqrt{\dfrac{2f_sL_m}{\eta R_{Lmin}}}$	$\sqrt{\dfrac{2f_sLn_1^2M_{VDCmax}^2}{\eta(1 - n_1M_{VDCmax})R_{Lmin}}}$
Max. magnetizing inductance $L_{m(max)}$	$\dfrac{n^2R_{Lmin}(1 - D_{Bmax})^2}{2f_s}$	$\dfrac{D_{min}V_{Imax}}{f_s\Delta i_{Lm(max)}}$
Max. inductance L_{max}	-not applicable-	$\dfrac{R_{Lmin}(1 - D_{Bmax})}{2f_s}$
Max. current ripple through the magnetizing inductance $\Delta i_{Lm(max)}$ (where $L_m < L_{m(max)}$)	$\dfrac{D_{min}V_{Imax}}{f_sL_m}$	$\alpha\Delta i_{Lmax}$ [b]
Max. current ripple through output filter inductance Δi_{Lmax} (where $L > L_{min}$)	-not applicable-	$\dfrac{(V_{Imax} - n_1V_O)D_{min}}{n_1f_sL}$

[a]where n_1 is the turns ratio between the primary and secondary winding and n_2 is the turns ratio between primary winding and tertiary (reset) winding.
[b]where α is a constant number less than 1.

Table B.3   Steady-state design equations for the flyback and forward dc–dc converters in DCM

Parameter/Topology	Flyback	Forward
Max. ESR of filter capacitor r_{Cmax}	$\dfrac{V_r}{2\Delta i_{Lm(max)}}$	$\dfrac{V_r}{\Delta i_{Lmax}}$
Min. filter capacitance C_{min} (where $r_C < r_{Cmax}$)	$\dfrac{2D_{max}V_O}{f_s R_{Lmin} V_r}$	$\max\left[\dfrac{D_{max}}{2f_s r_C}, \dfrac{1-D_{min}}{2f_s r_C}\right]$
Max. MOSFET voltage stress V_{SM}	$V_{Imax} + nV_O$	$\left(\dfrac{n_1}{n_3}+1\right)V_{Imax}$
Max. diode voltage stress V_{DM}	$\dfrac{V_{Imax}}{n} + V_O$	$V_{D1max} = \dfrac{V_{Imax}}{n_3}$ $\quad$ $V_{D2max} = \dfrac{V_{Imax}}{n_1}$ $\quad$ $V_{D3max} = \left(\dfrac{n_3}{n_1}+1\right)V_{Imax}$
Max. MOSFET current stress I_{SM}	Δi_{Lmax}	$I_{Imax} + \Delta i_{Lm(max)}$
Max. diode current stress I_{DM} $I_{D3max} = \left(\dfrac{n_1}{n_3}\right)\Delta i_{Lm(max)}$	$n\Delta i_{Lmax}$	$I_{D1max} = I_{D2max} = \Delta i_{Lmax}$

C

Simulation Tools

SPICE Model of Power MOSFETs

Figure C.1 shows a SPICE large-signal model for n-channel enhancement MOSFETs. It is a model of integrated MOSFETs, which can be adopted to power MOSFETs. SPICE parameters of the large-signal model of enhancement-type n-channel MOSFETs are given in Table C.1. The diode currents are

$$i_{BD} = IS \left(e^{\frac{v_{BD}}{V_T}} - 1 \right) \tag{C.1}$$

and

$$i_{BS} = IS \left(e^{\frac{v_{BS}}{V_T}} - 1 \right). \tag{C.2}$$

The junction capacitances in the voltage range close to zero are

$$C_{BD} = \frac{(CJ)(AD)}{\left(1 - \frac{v_{BD}}{PB} \right)^{MJ}} \quad \text{for } v_{BD} \le (FC)(PB) \tag{C.3}$$

and

$$C_{BS} = \frac{(CJ)(AS)}{\left(1 - \frac{v_{BS}}{PB} \right)^{MJ}} \quad \text{for } v_{BS} \le (FC)(PB) \tag{C.4}$$

where CJ is the zero-bias junction capacitance per unit area, AD is the drain area, AS is the source area, PB is the built-in potential, and MJ is the grading coefficient.

Laboratory Manual for Pulse-Width Modulated DC–DC Power Converters, First Edition.
Marian K. Kazimierczuk and Agasthya Ayachit.
© 2016 John Wiley & Sons, Ltd. Published 2016 by John Wiley & Sons, Ltd.

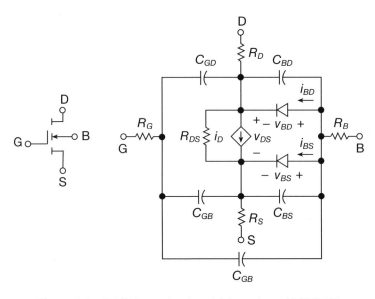

Figure C.1   SPICE large-signal model for n-channel MOSFET.

Table C.1   Selected SPICE Level 1 NMOS large-signal model parameters

Sym.	SPICE S.	Model parameter	Default value	Typical value
V_{to}	VTO	Zero-bias threshold voltage	0 V	0.3–3 V
μC_{ox}	KP	Process constant	$2 \times 10^{-5} A/V^2$	20–346 $\mu A/V^2$
λ	Lambda	Channel-length modulation	0 V^{-1}	0.5–10^{-5} V^{-1}
γ	Gamma	Body-effect V_t parameter	$0 \text{ V}^{\frac{1}{2}}$	$0.35 \text{ V}^{\frac{1}{2}}$
$2\phi_F$	PHI	Surface potential	0.6 V	0.7 V
R_D	RD	Drain series resistance	0 Ω	0.2 Ω
R_S	RS	Source series resistance	0 Ω	0.1 Ω
R_G	RG	Gate series resistance	0 Ω	1 Ω
R_B	RB	Body series resistance	0 Ω	1 Ω
R_{DS}	RDS	Drain-source shunt R	∞	1 MΩ
R_{SH}	RSH	Drain-source diffusion sheet R	0	20 Ω/Sq.m
I_S	IS	Saturation current	10^{-14} A	$10^{-9} A$
M_j	MJ	Grading coefficient	0.5	0.36
C_{j0}	CJ	Zero-bias bulk junction C/m^2	0 F/m^2	1 nF/m^2
V_{bi}	PB	Junction potential	1 V	0.72 V
M_{jsw}	MJSW	Grading coefficient	0.333	0.12
C_{j0sw}	CJSW	Zero-bias junction perimeter C/m	0 F/m	380 pF/m
V_{BSW}	PBSW	Junction sidewell potential	1 V	0.42 V
C_{GDO}	CGDO	Gate-drain overlap C/m	0 F/m	220 pF/m
C_{GSO}	CGSO	Gate-source overlap C/m	0 F/m	220 pF/m
C_{GBO}	CGBO	Gate-bulk overlap C/m	0 F/m	700 pF/m
F_C	FC	Forward-biased C_J coefficient	0.5	0.5
t_{ox}	TOX	Oxide thickness	∞	4.1 to 100 nm
μ_{ns}	UO	Surface mobility	600 cm^2/Vs	600 cm^2/Vs
n_{sub}	NSUB	Substrate doping	0 cm^{-3}/Vs	0 cm^{-3}/Vs

The junction capacitances in the voltage range far from zero are

$$C_{BD} = \frac{(CJ)(AD)}{(1 - FC)^{1+MJ}} \left[1 - (1 + MJ)FC + MJ\frac{v_{BD}}{PB} \right] \quad \text{for} \quad v_{BD} \geq (FC)(PB) \quad \text{(C.5)}$$

and

$$C_{BS} = \frac{(CJ)(AS)}{(1 - FC)^{1+MJ}} \left[1 - (1 + MJ)FC + MJ\frac{v_{BD}}{PB} \right] \quad \text{for} \quad v_{BS} \geq (FC)(PB). \quad \text{(C.6)}$$

The typical values:
$C_{ox} = 3.45 \times 10^{-5} \text{pF/}\mu\text{m}$
$t_{ox} = 4.1 \times 10^{-3} \mu\text{m}$
$\epsilon_{ox(SiO_2)} = 3.9\epsilon_0$
$C_{j0} = 2 \times 10^{-4} \text{F/m}^2$
$C_{jsw} = 10^{-9} \text{F/m}$
$C_{GBO} = 2 \times 10^{-10} \text{F/m}$
$C_{GDO} = C_{GSO} = 4 \times 10^{-11} \text{F/m}$

SPICE NMOS Syntax
```
Mxxxx D G S B MOS-model-name L = xxx  W = yyy
```

Example:
```
M1 2 1 0 0 M1-FET L = 0.18um W = 1800um
```

SPICE NMOS Model Syntax
```
.model model-name NMOS (parameter=value ...)
```

Example:
```
.model M1-FET NMOS (Vto = 1V Kp = E-4)
```

SPICE PMOS Model Syntax
```
.model model-name PMOS (parameter=value ...)
```

SPICE Subcircuit Model Syntax
```
xname N1 N2 N3 model-name
```

Example:
```
x1 2 1 0 IRF840
```

Copy and paste the obtained device model.
```
.SUBCKT  IRF840 1 2 3
```

and the content of the model.

Introduction to SPICE

SPICE is an abbreviation for *Simulation Program for Integrated Circuits Emphasis*. PSPICE is the PC version of SPICE. Analog and digital electronic circuit designs are verified widely by both industries and the academia using PSPICE. It is used to predict the circuit behavior.

Passive Components: Resistors, Capacitors, and Inductors

```
Rname N+  N- Value [IC = TC1]
Lname N+  N- Value [IC = Initial Voltage Condition]
Cname N+  N- Value [IC = Initial Current Condition]
```

Examples:

```
R1 1 2 10K
L2 2 3 2M
C3 3 4 100P
```

Transformer

```
Lp Np+ Np- Lpvalue
Ls Ns+ Ns- Lsvalue
Kname Lp Ls Kvcalue
```

Example:

```
Lp 1 0 1mH
Ls 2 4 100uH
Kt Lp Ls 0.999
```

Temperature

```
.TEMP list of temperatures
```

Example:

```
.TEMP 27 100 150
```

Independent DC Sources

```
Vname N+ N- DC Value
%Vname N$^{+}$ N$^{-}$ DC Value

Iname N+ N- DC Value
%Iname N$^{+}$ N$^{-}$ Type Value
```

Examples:

```
Vin  1 0 DC 10
Is 1 0 DC 2
```

DC Sweep Analysis

```
.DC Vsource-name Vstart Vstop Vstep
```

Example:

```
.DC VD 0 0.75 1m
```

Independent Pulse Source for Transient Analysis

```
Vname N+ N- PULSE (VL VH td tr tf PW T)
```

Example:

```
VGS 1 0 PULSE(0 1E-6 0 1 1 10E-6 100e-6)
```

Transient Analysis

```
.TRAN time-step time-stop
```

Example:
```
.TRAN 0.1ms  100ms 0ms 0.2ms
```

Independent AC Sources for Frequency Response

```
Vname N+ N- AC Vm Phase
Iname N+ N- AC Im Phase
```

Example:
```
Vs 2  3 AC 2 30
Is 2  3 AC 0.5 30
```

Independent Sinusoidal AC Sources for Transient Analysis

```
Vname N+ N- SIN (Voffset Vm f T-delay Damping-Factor Phase-delay)
Iname N+ N- SIN (Ioffset Im f T-delay Damping-Factor Phase-delay)
```

Examples:
```
Vin  1 0 SIN (0 170 60 0 -120)
Is 1 0 SIN (0 2 120 0 45)
```

AC Frequency Analysis

```
.AC DEC points-per-decade fstart fstop
```

Example:
```
.AC DEC 100 20 20
```

Operating Point

```
.OP
```

Starting the SPICE Program

1. Open the PSpice A/D Lite window (**Start** > **Programs** > **Orcad9.2 Lite Edition** > **PSpice AD Lite**).
2. Create a new text file (**File** > **New** > **Text File**).
3. Type the example code.
4. Save the file as fn.cir (e.g., Lab1.cir), file type: all files, and simulate by pressing the appropriate icon.
5. To include the Spice code of a commercial device model, visit the web site, for example, http://www.irf.com, http:www.onsemi.com, or http://www.cree.com. For example, for IRF devices, click on (**Design** > **Support** > **Models** > **Spice Library**).

Example Program: Diode I–V Characteristics

```
VD  1 0 DC 0.75V
D1N4001 1 0 Power-Diode
.model Power-Diode D (Is=195pA n=1.5)
```

```
.DC VD 0V 0.75V 1mV
.TEMP 27C 50C 100C 150C
.probe
\ins{                                    }.end
```

Introduction to MATLAB®

MATLAB® is an abbreviation for MATrix LABoratory. It is a very powerful mathematical tool used to perform numerical computation using matrices and vectors to obtain two- and three-dimensional graphs. MATLAB® can also be used to perform complex mathematical analysis.

Getting Started

1. Open MATLAB® by clicking **Start** > **Programs** > **MATLAB®** > **R2014a** > **MATLAB® R2014a**.
2. Open a new M-file by clicking **File** > **New** > **M-File**.
3. Type the code in the M-File.
4. Save the file as fn.m (e.g., Lab1.m).
5. Simulate the code by doing one of the following:
   - Click on **Debug** > **Run**.
   - Press F5.
   - On the tool bar, click the icon **Run**.
   Use **HELP** by pressing F1.
   Use % at the beginning of a line for comments.

Generating an *x*-axis Data
x=Initial-Value: Increment: Final-Value;

Example:
```
x=1:0.001:5;
```

or
x=[list of all the values];

Example:
```
x = [1, 2, 3, 5, 7, 10];
```

or
x = linspace(start-value, stop-value, number-of-points);

Example:
```
x = linspace(0, 2*pi, 90);
```

or x = logspace(start-power, stop-power, number-of-points);

Example:
```
x = logspace(1, 5, 1000);
```

Semilogarithmic Scale

```
semilogx(x-variable, y-variable);
semilogy(x-variable, y-variable);
grid on
```

Log–log Scale

```
loglog(x, y);
grid on
```

Generate an *y*-axis Data

```
y = f(x);
```

Example:

```
y = cos(x);

z = sin(x);
```

Multiplication and Division

A dot should be used in front of the operator for matrix multiplication and division.

```
c = a.*b;
c = a./b;
\end {verbatim}

\vskip 10pt
\noindent

{\bf Symbols and Units}

Math symbols should be in italic.
Math signs (like ( ), =, and +) and units should not be in italic.
Leave one space between a symbol and a unit.

\vskip 10pt
\noindent
{\bf $x$-axis and $y$-axis Labels}

xlabel('\{$\backslash$it {x}\} (unit) ')

ylabel('\{$\backslash$it {y}\} (unit) ')


Example:
\begin{verbatim}
xlabel('{\it  v_{GS}} (V)')
ylabel('{\it  i_{DS} } (A)')
```

x-axis and *y*-axis Limits

set(gca, 'xlim', [xmin, xmax])
set(gca, 'xtick', [xmin, step, xmax])
set(gca, 'xtick', [x1, x2, x3, x4, x5])

Example:

```
set(gca, 'xlim', [1, 10])
set(gca, 'xtick', [0:2:10])
set(gca, 'xtick', [-90 -60 -30 0 30 60 90])
```

Greek Symbols

Type: \alpha , \beta , \Omega , \omega , \pi , \phi , \psi , \gamma , \theta , and \circ
to obtain: $\alpha, \beta, \Omega, \omega, \pi, \phi, \psi, \gamma, \theta$, and $\circ$.

Plot Commands

plot (x, y, '.-', x, z, '- -') set(gca, 'xlim', [x1, x2]); set(gca, 'ylim', [y1, y2]); set(gca, 'xtick', [x1:scale-increment:x2]); text(x, y, '*symbol* = 25 V'); plot (x, y), axis equal

Examples:

```
set(gca, 'xlim', [4, 10]);
set(gca, 'ylim', [1, 8]);
set(gca, 'xtick', [4:1:10]);
text(x, y, '{\it V} = 25 V');
```

3D Plot Commands

[X1, Y1] = meshgrid(x1,x2);
mesh(X1, Y1,z1);

Example:

```
t = linspace(0, 9*pi);
xlabel('sin(t)')
ylabel('cos(t)')
zlabel('t')
plot(sin(t), cos(t), t)
```

Bode Plots

```
f = logspace(start-power, stop-power, number-of-points)

NumF = [a1 a2 a3];%~~~~~~~%Define the numerator of polynomial in
$s$-domain.
DenF = [a1 a2 a3];%~~~~~~~%Define the denominator of polynomial in
$s$-domain.
[MagF, PhaseF] = bode(NumF, DenF, (2*pi*f));
```

```
figure(1)
semilogx(f, 20*log10(MagF))
F = tf(NumF, DenF)%~~~~~~~\%Converts the polynomial into transfer
function.
[NumF, DenF] = tfdata(F)%~~~~~~~\%Converter transfer function into
polynomial.
```

Step Response
```
NumFS = D*NumF;
t = [0:0.000001:0.05];
[x, y] = step(NumFS, DenF, t);
figure(2)
plot(t, Initial-Value + y);
```

To Save Figure
Go to File, click Save as, go to EPS file option, type the file name, and click Save.

Example Program
```
clear all
clc
x = linspace(0, 2*pi, 90);
y = sin(x);
z = cos(x);
grid on
xlabel('{\it x}')
ylabel('{\it y }, {\it  z }')
plot(x , y, '-.', x, z, '-~-')
```

Polynomial Curve Fitting
```
x = [0 0.5 1.0 1.5 2.0 2.5 3.0];
y = [10 12 16 24 30 37 51];
p = polyfit(x, y, 2) %Find the coefficients of a polynomial of
degree 2
yc = polyval(p, x); %Polynomial is evaluates at all points.
plot(x, y, 'x', x, yc)
xlabel('x')
ylabel('y'), grid
legend('Actual data', 'Fitted polynomial')
```

Bessel Functions
J0 = besselj(0, x);

Modified Bessel Functions
I0 = besseli(0, x):

Example:
```
model = [1 2 3]:
rro = -1:0.00001:1;
```

```
kr = (1 + j)*(rodel)'*(rro);
JrJ0 = besseli(0,kr);
figure, plot(rro, abs(JrJ0))
figure, plot(rro,angle(JrJ0)*180/pi)
```

SABER Circuit Simulator

SABER is a circuit simulator. It enables circuit analysis, schematic design, mixed-mode circuit simulation, waveform analysis, and report generation.

Starting SABER

1. This procedure is specific to Wright State University. In the Start menu, locate and click "XWin32 2011." In the same way, find "Secure Shell" and click "Secure Shell Client." A terminal console will appear. Hit Enter. Enter "thor.cs.wright.edu" as host name, the UNIX log-on name as username and click OK. Enter your Password when prompted. Now, you are connected remotely.
2. Type "sketch" on the terminal console. SABER will start automatically.

Setting up a circuit on SABER

1. SABER created many files during netlisting and simulations. In order to keep everything in order, create a new folder in your home directory called "Saber" (or "Lab1", etc.). Use the SSH Secure File Transfer Client, or type "mkdir Saber" at the home directory command prompt.
2. On the SABER circuit design window, construct a given circuit/schematic using the parts in the component window. The parts can be found in the component library at the bottom icon tray on your screen under "Select and place parts," and you can search for parts individually. Left click at the terminal to connect your schematic. You can search for parts individually. NOTE: It is recommended to save your work regularly with a suitable file name. Make sure to save all your work in a designated folder in the Secure Shell Transfer Client because SABER generates netlists that are required for circuit simulations.
3. Provide appropriate values to all the components on the schematic and name the wires that are used to connect these components, such as Vout, Vin, iL, etc. This can be done by right clicking on the selected wire and then selecting "attribute."

Transient Analysis of a Circuit with SABER

1. Select the show/hide Saber Guide icon bar to view the simulation tool bar.
2. Select an Operating Point/Transient Analysis to run transient analysis versus time.
   - Click the Basic Tab.
   - Set a simulation end time. This time is different for each experiment. Assign appropriate enc times based on the time taken for the system to reach steady-state operation. Set a time step of $0.01\,\mu s$. If the end time is long, it is recommended to increase the time step so that the simulation runs faster.
   - Set Monitor Progress = **100**. Plot after Analysis=**Yes Open Only**.

- Click **Input/Output** Tab. Select **Input/Output** Tab. Select **All Signals for the Signal List**.
- Select **Across and through variables** in order to save both voltage and current information.
- Select OK to run the analysis. It will take 1 or 2 minutes to run it. If you want to view the status of the simulation, click the Simulation Transcript and Command Line button on the top bar.
- Once the analysis is completed, it should open Cosmos Scope to show the simulation results.

Plotting with SABER

1. The Cosmos Scope pops up with two more dialog boxes. In the Signal Manager window, under Plot files, you will see the file you have been working on, which contains the results of the transient analysis.
2. All types of waveforms can be plotted and various mathematical operations can be performed on them. You should be able to find the required waveforms under the drop-down option corresponding to the component under consideration.
3. Each lab will have different sets of waveforms to be plotted.

D

MOSFET Parameters

Part number	V_{DSS} (V)	I_{SM} (A)	r_{DS}	C_o (pF)	V_{th} (V)
IRF150	100	40	0.055 Ω	100	4
IRF430	500	4.5	1.8 Ω	135	4
NTD3055	60	12	94 mΩ	107	4
IRF142	100	24	0.11 Ω	100	4
IRF840	500	8	0.85 Ω	100	4
IRF740	400	10	0.55 Ω	100	4

Laboratory Manual for Pulse-Width Modulated DC–DC Power Converters, First Edition.
Marian K. Kazimierczuk and Agasthya Ayachit.
© 2016 John Wiley & Sons, Ltd. Published 2016 by John Wiley & Sons, Ltd.

E

Diode Parameters

Part number	V_{RRM} (V)	I_{Fave} (A)	R_F (mΩ)	V_F (V)
MBR1060	60	20	25	0.4
MUR1560	600	15	17.1	1.5
MBR10100	100	10	15	0.8
MBR10100	100	10	15	0.8
MUR2510	100	25	20	0.7
MUR2540	40	25	10	0.3
MR826	600	35	—	—
MBR4040	40	40	25	0.4

Laboratory Manual for Pulse-Width Modulated DC–DC Power Converters, First Edition.
Marian K. Kazimierczuk and Agasthya Ayachit.
© 2016 John Wiley & Sons, Ltd. Published 2016 by John Wiley & Sons, Ltd.

F

Selected MOSFETs Spice Models

IRF430

```
************************************************************************
.model IRF430   NMOS(Level=3 Gamma=0 Delta=0 Eta=0 Theta=0 Kappa=0.2 Vmax=0 Xj=0
+               Tox=100n Uo=600 Phi=.6 Rs=2.514m Kp=20.56u W=.33 L=2u Vto=3.783
+               Rd=.944 Rds=2.222MEG Cbd=677.2p Pb=.8 Mj=.5 Fc=.5 Cgso=1.725n
+               Cgdo=136.7p Rg=1.159 Is=46.85p N=1 Tt=585n)
************************************************************************
```

IRF520

```
************************************************************************
.model IRF520   NMOS(Level=3 Gamma=0 Delta=0 Eta=0 Theta=0 Kappa=0.2 Vmax=0 Xj=0
+               Tox=100n Uo=600 Phi=.6 Rs=.1459 Kp=20.79u W=.73 L=2u Vto=3.59
+               Rd=80.23m Rds=444.4K Cbd=622.1p Pb=.8 Mj=.5 Fc=.5 Cgso=517.9p
+               Cgdo=137.3p Rg=6.675 Is=2.438p N=1 Tt=137n)
************************************************************************
```

IRF150

```
************************************************************************
.model IRF150   NMOS(Level=3 Gamma=0 Delta=0 Eta=0 Theta=0 Kappa=0 Vmax=0 Xj=0
+               Tox=100n Uo=600 Phi=.6 Rs=1.624m Kp=50u W=160u L=2u Vto=2
+               Rd=1.031m Rds=444.4K Cbd=3.229n Pb=.8 Mj=.5 Fc=.5 Cgso=9.027n
+               Cgdo=1.679n Rg=13.89 Is=194E-18 N=1 Tt=288n lambda=0.05)
************************************************************************
```

IRF142

```
************************************************************************
.model IRF142   NMOS(Level=3 Gamma=0 Delta=0 Eta=0 Theta=0 Kappa=0.2 Vmax=0 Xj=0
+               Tox=100n Uo=600 Phi=.6 Rs=21.94m Kp=20.61u W=.97 L=2u Vto=3.189
+               Rd=42.19m Rds=444.4K Cbd=2.398n Pb=.8 Mj=.5 Fc=.5 Cgso=1.113n
+               Cgdo=432.1p Rg=5.659 Is=2.823p N=1 Tt=142n)
************************************************************************
```

Laboratory Manual for Pulse-Width Modulated DC–DC Power Converters, First Edition.
Marian K. Kazimierczuk and Agasthya Ayachit.
© 2016 John Wiley & Sons, Ltd. Published 2016 by John Wiley & Sons, Ltd.

IRF840

```
*************************************************************************
.model IRF840  NMOS(Level=3 Gamma=0 Delta=0 Eta=0 Theta=0 Kappa=0.2 Vmax=0 Xj=0
+              Tox=100n Uo=600 Phi=.6 Rs=6.382m Kp=20.85u W=.68 L=2u Vto=3.879
+              Rd=.6703 Rds=2.222MEG Cbd=1.415n Pb=.8 Mj=.5 Fc=.5 Cgso=1.625n
+              Cgdo=133.4p Rg=.6038 Is=56.03p N=1 Tt=710n)
*************************************************************************
```

IRF740

```
*************************************************************************
.model IRF740  NMOS(Level=3 Gamma=0 Delta=0 Eta=0 Theta=0 Kappa=0.2 Vmax=0 Xj=0
+              Tox=100n Uo=600 Phi=.6 Rs=8.563m Kp=20.59u W=.78 L=2u Vto=3.657
+              Rd=.3915 Rds=1.778MEG Cbd=1.419n Pb=.8 Mj=.5 Fc=.5 Cgso=1.392n
+              Cgdo=146.6p Rg=.9088 Is=17.65p N=1 Tt=570n)
*************************************************************************
```

G

Selected Diodes Spice Models

MUR1560

```
***************************************************************************
.MODEL Dmur1560 d
+IS=1.22946e-07 RS=0.0276435 N=2 EG=1.3 XTI=4 BV=600 IBV=0.00001
+CJO=5.43205e-10 VJ=0.4 M=0.373483 FC=0.5 TT=7.14134e-08 KF=0 AF=1
***************************************************************************
```

MBR10100

```
***************************************************************************
.MODEL Dmbr10100 d
+IS=0.000132385 RS=0.0122186 N=2 EG=0.828762 XTI=3.80757 BV=100 IBV=0.0001
+CJO=1e-11 VJ=0.7 M=0.5 FC=0.5 TT=0 KF=0 AF=1
***************************************************************************
```

MBR1060

```
***************************************************************************
.MODEL Dmbr1060 d
+IS=0.000132385 RS=0.0122186 N=2 EG=0.828762 XTI=3.80757 BV=60 IBV=0.0001
+CJO=1e-11 VJ=0.7 M=0.5 FC=0.5 TT=0 KF=0 AF=1
***************************************************************************
```

MUR2510

```
***************************************************************************
.MODEL MUR2510  D(Is=67.39p Rs=4.351m Ikf=0 N=1 Xti=3 Eg=1.11 Cjo=243.6p
+              M=.1593 Vj=.75 Fc=.5 Isr=100.8n Nr=2 Tt=123.3n)
***************************************************************************
```

Laboratory Manual for Pulse-Width Modulated DC–DC Power Converters, First Edition.
Marian K. Kazimierczuk and Agasthya Ayachit.
© 2016 John Wiley & Sons, Ltd. Published 2016 by John Wiley & Sons, Ltd.

MBR2540

```
***************************************************************************
.model MBR2540  D(Is=21u Rs=5.011m Ikf=69.6 N=1 Xti=0 Eg=1.11 Cjo=3.507n
+               M=.5031 Vj=.75 Fc=.5 Isr=1.176m Nr=2)
***************************************************************************
```

MBR4040

```
***************************************************************************
.model MBR4040  D(Is=27.51u Rs=5.244m Ikf=124.2 N=1 Xti=0 Eg=1.11 Cjo=3.468n
+               M=.5003 Vj=.75 Fc=.5 Isr=1.175m Nr=2)
***************************************************************************
```

H

Physical Constants

Physical Constants and Values of Semiconductor Material Properties

This appendix provides values of the physical constants and the values of the physical properties of silicon (Si), silicon-carbide (SiC), and gallium-nitride (GaN) semiconductor materials, which are helpful in solving the activities in Part III of this manual.

Table H.1 provides the values of physical properties. Table H.2 lists the values of all the critical physical properties of Si, SiC, and GaN semiconductor materials.

Table H.1 Physical constants and their values

Property	Symbol	Unit	Value
Thermal voltage	$V_T = \frac{kT}{q}$	V	0.0259 at $T = 300$ K
Boltzmann's constant	$k = \frac{E_T}{T}$	J/K	$1.3806488 \times 10^{-23}$
Boltzmann's constant	$k = \frac{E_T}{T}$	eV/K	8.62×10^{-5}
Planck's constant	h	J·s	6.62617×10^{-34}
Planck's constant	h	eV·s	4.14×10^{-15}
Magnitude of electron charge	q	C	1.60218×10^{-19}
Free-space permittivity	$\epsilon_0 = \frac{10^{-9}}{36\pi}$	F/m	8.85418×10^{-12}
Free-space permeability	μ_0	H/m	$4\pi \times 10^{-7}$
Speed of light in free space	$c = \frac{1}{\sqrt{\epsilon_0\mu_0}}$	m/s	2.998×10^8
Mass of free electron	m_e	kg	9.31×10^{-31}
Mass of free hole	m_h	kg	1.673×10^{-27}
Intrinsic concentration of Si at $T = 300$ K	n_i	cm^{-3}	1.0987×10^{-8}
Maximum junction temperature of Si	T_{Jmax}	°C	600
Thermal conductivity of Si	k_{th}	W/K·cm	4.56

Laboratory Manual for Pulse-Width Modulated DC–DC Power Converters, First Edition.
Marian K. Kazimierczuk and Agasthya Ayachit.
© 2016 John Wiley & Sons, Ltd. Published 2016 by John Wiley & Sons, Ltd.

Table H.2   Properties of silicon, silicon carbide, and gallium nitride

Property	Symbol	Unit	Value
Silicon band gap energy	$E_{G(Si)}$	eV	1.12
Silicon band gap energy	$E_{G(Si)}$	J	1.793×10^{-19}
Silicon-carbide band gap energy	$E_{G(SiC)}$	eV	3.26
Silicon-carbide band gap energy	$E_{G(SiC)}$	J	5.216×10^{-19}
Gallium-nitride band gap energy	$E_{G(GaN)}$	eV	3.39
Gallium-nitride band gap energy	$E_{G(GaN)}$	J	5.430×10^{-19}
Silicon-dioxide band gap energy	$E_{G(SiO_2)}$	eV	9
Silicon-dioxide band gap energy	$E_{G(SiO_2)}$	J	14.449×10^{-19}
Silicon breakdown electric field	$E_{BD(Si)}$	V/cm	2×10^5
Silicon-carbide breakdown electric field	$E_{BD(SiC)}$	V/cm	22×10^5
Gallium-nitride breakdown electric field	$E_{BD(GaN)}$	V/cm	33×10^5
Silicon-dioxide breakdown electric field	$E_{BD(SiO_2)}$	V/cm	60×10^5
Silicon relative permittivity	$\epsilon_{r(Si)}$	–	11.7
Silicon-carbide relative permittivity	$\epsilon_{r(SiC)}$	–	9.7
Gallium-nitride relative permittivity	$\epsilon_{r(GaN)}$	–	8.9
Silicon-dioxide relative permittivity	$\epsilon_{r(SiO_2)}$	–	3.9
Silicon electron mobility at $T = 300$ K	$\mu_{n(Si)}$	cm^2/V·s	1360
Silicon-carbide electron mobility at $T = 300$ K	$\mu_{n(SiC)}$	cm^2/V·s	900
Gallium-nitride electron mobility at $T = 300$ K	$\mu_{n(GaN)}$	cm^2/V·s	2000
Silicon hole mobility at $T = 300$ K	$\mu_{p(Si)}$	cm^2/V·s	480
Silicon-carbide hole mobility at $T = 300$ K	$\mu_{p(SiC)}$	cm^2/V·s	120
Gallium-nitride hole mobility at $T = 300$ K	$\mu_{p(GaN)}$	cm^2/V·s	30
Silicon effective electron mass coefficient	k_e	–	0.26
Silicon effective hole mass coefficient	k_h	–	0.39
Silicon-carbide effective electron mass coefficient	k_e	–	0.36
Silicon effective hole mass coefficient	k_h	–	1
Gallium-nitride effective electron mass coefficient	k_e	–	0.23
Gallium-nitride effective hole mass coefficient	k_h	–	0.24

I

Format of Lab Report

1. All reports must be written using Microsoft Word or LaTeX in a single-column format.
2. Make sure you write the equations using the Microsoft Word Equation feature available under the Insert tab. The font size must not exceed 12 and must not be less than 10. Using Times New Roman, Calibri, or Cambria as font styles is recommended.
3. Make sure you include the cover page at the beginning of every lab report in order to distinguish between different labs and different teaching assistants. The cover page has been made available on Pilot.
4. All figures, those used to illustrate your ideas must be drawn legibly either using Microsoft Visio or any other drawing tool. You may choose to draw figures, if need be, by hand, but they must have good clarity.
5. All MATLAB® codes must be included in the Appendix section at the end of your report.
6. Your lab reports for all the labs must consist of the sections mentioned below. Failure to follow the instructions will result in a poor lab grade.

I. ABSTRACT

In this section, mention in a few lines the gist of your experiment. You should indicate what aspects you will cover, how you would plan to arrive at the analysis, and the resources used to perform the analysis. You may add other relevant details that you deem appropriate in this section.

II. INTRODUCTION

In this section, you introduce to the readers the concept under consideration. You may discuss the operation of the circuits, laws, or any theory you will be working on. You may define some terms, write expressions, and build sufficient theory to support your work. The different terms that you will be using throughout the report can be introduced here to give the reader an idea of what you are going to achieve. You are encouraged to refer to various articles and documents available

Laboratory Manual for Pulse-Width Modulated DC–DC Power Converters, First Edition.
Marian K. Kazimierczuk and Agasthya Ayachit.
© 2016 John Wiley & Sons, Ltd. Published 2016 by John Wiley & Sons, Ltd.

on the internet to provide good details about your work. However, you have to cite the resource you have used to give proper credit to the original or true authors of the source.

III. ANALYSIS

Here, you will include all the relevant details pertaining to the experiment. You will include the equations, calculations, or any other math work that you consider important. All the values must be represented with proper units and providing equation numbers is encouraged to improve the clarity. You will also discuss in short the procedure that you used to arrive at the results. Please do not include any results or conclusion in this section.

IV. RESULTS

In this section, you will discuss the results of the various plots, simulations, or calculations that you would have performed as a part of the lab requirement. All your plots have to adhere to IEEE format only. A sample plot and the code used to generate the sample plot are given below. Careful attention needs to be paid to providing units for all the variables, and they must be readable. For figures with multiple subfigures, you must use different line types to distinguish between different results. If there are any values that you obtained from MATLAB®, summarize them in a well-organized table. Please do not copy all the results from the command window or the command history on MATLAB® and paste them in your work.

A sample plot which follows the IEEE format is shown in the figure below. This plot is similar to the variation in intrinsic carrier concentration with respect to change in temperature.

A snippet of the MATLAB® code used to output the figures in IEEE format is as given below:

```
figure(1)
semilogy(T, Si_cm, T, ni_SiC_cm)
xlabel('{\it T} (K)', 'FontSize', 12)
ylabel('{\it n_i} (cm^{-3})', 'FontSize', 12)
grid on
text(500, 10^(7), '{\it n_i}_{({SiC})}', 'FontSize', 12)
text(350, 5*10^(13), '{\it n_i}_{({Si})}', 'FontSize', 12)
ylim([10^(-20) 10^(20)])
```

Simulation Results: If you have any simulation results performed on SABER, LTSpice, or Spice, which is relevant to your experiment, make sure you consider the following details:

- Your plots must appear on a white background. The thickness of the waveforms must be increased.
- If you have more than three waveforms, plot them on different figure windows.
- Make sure each waveform is plotted using a different color or line style.

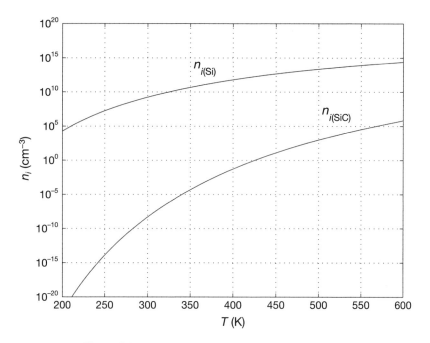

Figure I.1   Sample plot representing the IEEE format.

- Mention the variable name if possible.
- Mention your name on every plot.

V. CONCLUSION

In this section, you will summarize all your work and describe your results concisely. You may include numbers, theoretical validations, and simulation observations as a part of your conclusion.

VI. POSTLAB

Post-lab questions asked at the end of the lab handout will be answered in this section. Your answers must be well-organized and clearly indicate the question numbers. You are encouraged to justify your reasoning using any material available on the internet and proper citation must be provided for the same.

VII. REFERENCES

References are a must. No work can be performed without the help of references. Remember that your class notes or the reference textbook can also be included as references.

VIII. APPENDIX

Your MATLAB® code goes here. You may copy and paste the editor window results directly on to this section. Please do not print.

Index

Laboratory Manual for Pulse-Width Modulated DC–DC Power Converters, First Edition.
Marian K. Kazimierczuk and Agasthya Ayachit.
© 2016 John Wiley & Sons, Ltd. Published 2016 by John Wiley & Sons, Ltd.